Preface

Throughout this book you will find the frequent use of the word ***IT***, which will be in bold, capitalized italic letters, as a neutral pronoun. The meaning of ***IT*** unfolds with the progression of each section, yet until the end, ***IT*** may not be altogether clear, so until then, consider ***IT*** as a new way to say God, Creator, or even "pure energy." ***IT*** signifies that which put and keeps this Universe together.

Additionally, this book contains information intended to challenge your current understanding of the universe and its workings along with our existence within ***IT***. The topics covered may be familiar, and they are backed by science or personal experience, yet it is unlikely that you have viewed our existence from the perspective of Omnipresent. To fully understand the atom, for example, or to appreciate the production of garbage or the reshaping of ***IT***. You would benefit from preparing for this book as you would a meditation - by unlocking the mind.

To assist you in unlocking your mind I have included the photo that follows. Look at it, and think about what you see.

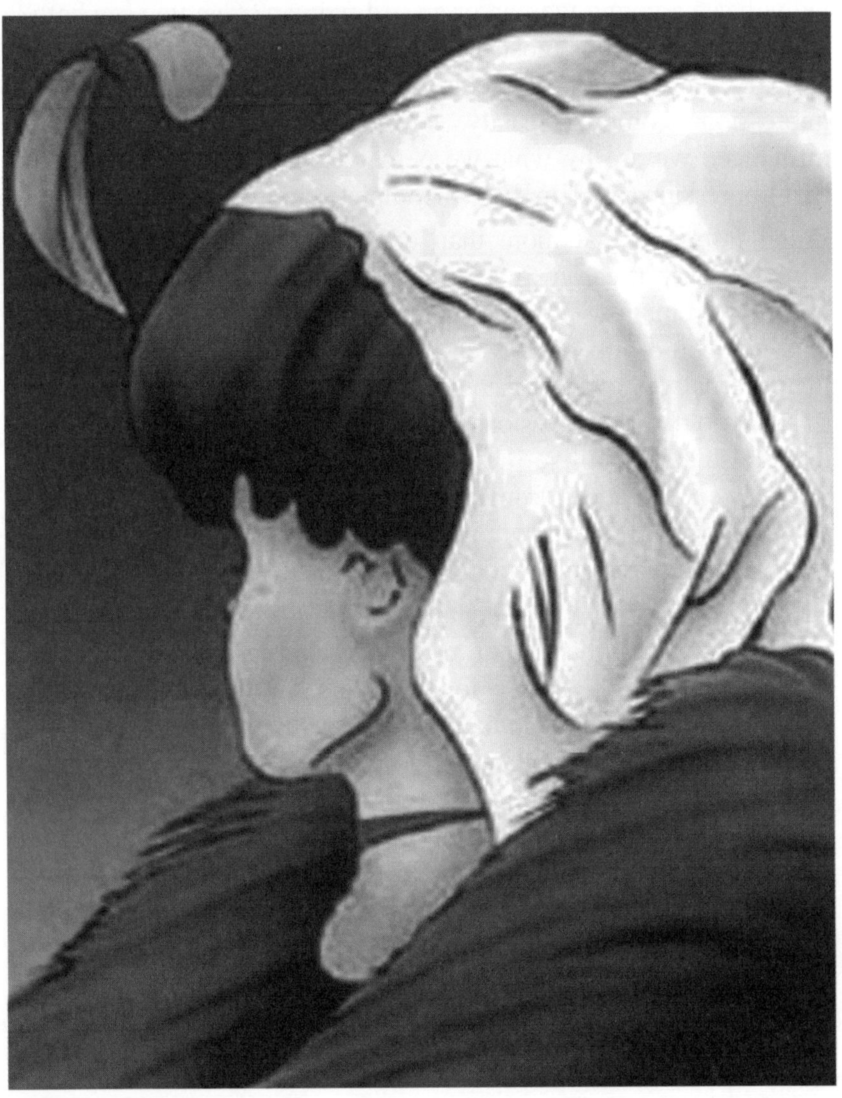

[1]This artwork is in the public domain. It was first publish as an anonymous German postcard ca. 1888. It was redone for an advertisement for the Anchor Buggy Company from 1890. Thereafter, British cartoonist W. E. Hill published it in 1915 in *Puck* humor magazine, an American magazine inspired by the British magazine *Punch*.

Do you see a young woman or an old woman? Can you see both? Both are there! For most of us, once our mind locks in one view, usually the first, it will seldom look for alternatives. Our first impressions come so easily, so quickly, yet considering how influenced we are by what we perceive, we should be aware that first impressions are limited, incomplete, and very often incorrect.

Being informed that more than one figure in the picture exists likely makes it easier to see the image differently. However, the question remains: Without knowing an opposite alternative existed, would you have even tried to see it? By first accepting that alternatives exist, you begin to unlock your mind, clearing it of all the first impressions enabling you to see the other image that is also there. Subsequently, when we take the opportunity to view even old ideas from new angles, we will find those quick and easy ways of seeing are not the correct or best methods to use when engaging our minds in a new thought.

Section **one** of Omnipresent sets forth what I call trivia: a collection of thoughts meant to provide alternative ways of seeing and being. It's a primer for sections two and three, which uncover questions and truths about time, matter, and the mass of the universe, about God, and ***IT*** as an ever changing place of omnipresence. I hope you will enjoy the book. How I came upon the material within is a story in itself, but in summary I can say it all stems from gratitude. I am grateful that I exist and that I accept and question what I see. In doing so, I have seen as most of us see. And because of my gratitude I have learned to see alternatives, which has helped me in a personally profound way to know who I really am.

PART #1

Trivia for Unlocking the Mind

A moment before and after the Big Bang, the mind did not exist. Only ***IT*** as pure energy existed. Only in the process of ***IT*** rearranging ***ITSELF*** as this pure energy did it form what we know as the human mind.

As often as I try to forget who I think I am, it hits me in the face every day that my mind sees me in the mirror cleaning and tightening this skeleton surrounded by water that I call my self. It is in fact due to this water shell surrounding the mind that it is hard for most of us to detach from who we think we are. We think we are merely the image in the mirror: skin, hair, a body: overweight, underweight, fit or arthritic. We do not know ourselves as pure energy, especially not the same pure energy that existed the moment before and after the Big Bang.

From the moment we are born, we gather our knowledge from what we see and hear in our environment. To make sense of life, we develop theories and maxims such as seeing is believing. Yet, we do not see all that exists. We see grass only as green, for example, when it is every color of the spectrum except the color green. Our eyes only detect the color reflected by the object, so our minds exclude from our consciousness all the colors being absorbed by the object, yet these colors are also present.

We also believe that whatever goes up must come down. This is one of the concepts that I learned when I was young. But not everything sent into space has come back to the Earth. Take the spacecraft Voyager; it will not come back down.

Another thing I learned came from the slogan: “Only you can prevent forest fires." But, the first forest fires and many since have started from lightning.

I did not create lightning, nor can I prevent it. Nonetheless, we can thank lightning for man learning about fire and enjoying cooked meat just as we can thank GOD for the existence of lightning and meat. We should actually be grateful to GOD everyday, even multiple times a day if not for the whole day.

In certain forms of religion the mind practices accepting GOD on certain days of the week. But accepting of GOD once or weekly does not guarantee this understanding will never be lost. It is very possible to lose it.

The mind, which has no difficulty accepting its own existence or the existence of other minds, struggles to accept there is more to be grateful for than what the mind can control. The mind has trouble believing in the existence of God, the pure energy that I call ***"IT."***

❁~~~~~~~~~~~~~~~~~~~~~❁~~~~~~~~~~~~~~~~~~~~~❁

*** ***The human mind is made in ITS image. Like GOD, the mind does not want to be told what to do or to be closed down. Like the mind, IT wants to be recognized.*** ***

⌘~~~~~~~~~~~~~~~⌘~~~~~~~~~~~~~~~⌘~~~~~~~~~~~~~~~⌘

The mind does not have trouble talking about these things—talking keeps the mind active. However, the moment the mind is pushed to accept GOD as omnipresent pure energy, conflict arises. When I first confronted the idea of omnipresence, my mind led me to my environment for answers. But there were no answers.

The mind is conditioned to believe only what it can identify, name, define, and manipulate. Additionally, the mind invites manipulation as it picks up information that it believes to be important. This has been useful to advertisers and the marketplace. We've all likely seen products priced at $9.99. The mind sees this as a deal because nine dollars is less than ten dollars, even if only by a cent.

The federal government may eventually discontinue the penny, when it could costs more than its value to manufacture.

Then we will see prices increase by a cent. Will we feel ripped off?

Here is another example: Buy one, get one at 50% off. Why not simply put a 25% discount on any one pair of shoes? That is what the buy one get the second one half off sale is essentially offering. But instead, we are manipulated to believe it is a deal to buy an extra, probably unneeded pair of shoes.

Advertisers aren't the only people that try to dupe us. People that call themselves friends even if unintentionally hurt us under the guise of help. Let's say you lose your job and you schedule a garage sale to make some extra money. Some of your friends come to the garage sale to help you. One sees that you have a radio for sale for $20.00 and he offers you $10.00—because he is your friend. You would think that if he were a real friend to be a real help, he would offer you $30.00.

But we believe in good intentions be they of advertisers and misguided friends, just as we believe the words of televised news media. We accept as true a reporter's claim that "Everyone attended the meeting" or "The whole country grieves over the death of Mr. X." But these statements are exaggerations, even lies. Not everyone can be in one place at one time, nor will everyone ever agree or feel the same regarding a single event or experience.

A holy or righteous person might find being alone, such as in meditation with the pure energy of the inner self, a glorious experience while another person might find this same conscious solitude with the internal creative energy a torture. But, the real torture here is how the mind manipulates the significance of the events to satisfy its own need for control.

How will you know when it's your inner self and not your mind in control?

The mind categorizes everything. The mind devised the word "history," for example, in order to categorize the events of the past and promote a feeling that it has some control over them. History is thus what the human mind uses to confirm its existence according to the many things humans have done on this planet.

❀~~~~~~~~~~~~~~~~~~~~~~~❀❀~~~~~~~~~~~~~~~~~~~~~~❀

****The mind gave everything a name that has a name attached to it. *****

⌘~~~~~~~~~~~~~~~~~~~~~~~⌘~~~~~~~~~~~~~~~~~~~~~~⌘

The human mind also loves to worry. Worrying helps the mind to stay active. But as the mind resists being disconnected from its environment, worrying actually distracts us from our environment. It keeps our inner selves from smelling the roses.

Knowing that the mind prefers the controlling activity of being in the driver's seat, it's understandable why we worry more than we relax or meditate. During meditation the mind must ride in the back seat so to speak.

Most of us have experienced a time of insistent worry when we were ready to go to sleep. The mind brings up problems that have nothing to do with the existing moment. Then when we finally do fall asleep, the mind uses the information it has accumulated through worrying to continue the mind's activity in the form of dreams. Notice that when you are waking up, the thoughts you had while dreaming linger in your mind.

Then when you wake completely from the dream state, your mind will return to the worries it went to sleep with. Meditation can thus provide even more rest for the mind than sleep can.

The mind also has wants. When all you hear within your head is "I want this," I want that," "Do it this way, or that way," etc., your mind is dominating over your inner self; it's manipulating you. The mind thinks it will be happier or sleep better with a million dollars in the bank. And we work hard to get what we want even when wanting actually limits what we attain in life. It inhibits what we can learn about ourselves and what we are to do here, for when we want, we aren't listening to ***IT***. We say we would prefer to work less. We say we would prefer more freedom. But less work and more freedom means diminishing our other wants. How badly do you want to understand the saying:" Be careful for what you ask for; you just may get it"?

Who is smarter, I or the mind?

Here is one that I too have had to question, as to who is smarter, I as how I think I am, or that inner I that is there as a very powerful force. My personal feeling is that there is two of me.

There is the me that is conscious of myself, which I find deals with my outside environment, and let me add that I do talk to myself, and I have asked myself who am I talking to when I talk to myself. I get the feeling that there is a place inside of me where I can just talk to myself, where there is no danger, for I do not seem to get lost inside of me when I have talked to myself. I do feel that I am talking to me, as my mind which has the extraordinary ability to be able to talk and listen at the same moment.

And then there is that other one inside me, and I know it exists, for when I know that I should not eat certain things, like sweets, because sweets to me are an extra pound of gained weight, for I know that after I finish that whole cheesecake I can in a few days weigh myself again and I will weigh one pound more, and yes I do constantly, like every few days, check my weight. I have noticed that if I stick to my disciplinary diet I can control my weight, and that if I eat sweets just for one day my weight does not change. But when I eat, let us say sweets from bananas continually, because I bought too many, I can see the weight go up in a few days. Now I reduce the act of buying sweets, and I then eat all the sweets that Sizzler has, and I have also noticed that after my banquet at "all you can eat" at Sizzler, if I return to my disciplinary diet I do not gain any weight from that one day of feasting.

But let me return to that other inner self that I have to deal with, for as the above, I know that I have to tell myself, or that one that exists inside of me, that if we stick to my regular diet we, as the one that is inside of me that likes to gain weight, can then also enjoy sweets on certain occasions.

So I have given you things that I go through and that some of you also go through. Let me continue using the cheesecake as an example. This cheesecake that I just opened to cut out one small piece becomes a problem because then that other one that is inside of me will say "go ahead, have one more." It becomes a battle as I want to stop, and my inner other wants to finish the whole cake. I know that all I wanted was just a taste of the cheesecake, but it becomes a problem when I know that I have to constantly control my weight, or rather my mouth, for at the moment that I finish that one piece my inner other kicks in and I become aware of its existence. I only wanted one small piece, and now I am in battle with that other who wants to have one more so that I can get fat, when I knowingly only wanted one piece of the cheesecake.

So, I end up saying to that other me that I should not eat more. Well, as some of you have also found out, it seems that the other can also fight a good battle. In order to have the proper conditions to fight this other I found that the best thing is not to fuel it by buying or doing any of the things that it likes. To me that would be getting sweets, to other people it could be other things.

I have to be very alert because I know that there is that other part of me that has exercised its power over me. For example, as an alcoholic I have to discipline myself not to consume any alcohol. The bottom line is that I know that discipline works.

I feel that the other one inside of me is always there ready to pounce and make me do something stupid. If anyone out there knows how to lock this other inside us somewhere where one can throw away the key, let me know. I would love to be free to live my life without this constant battle.

Let me also add that whoever this other is, it was born with me. I have kept this other one that I exist as more in control when it comes to my outside activities, which are governed more by certain disciplines. By that I mean that when I have to do something as an existing moment I do not question that other in me, I just do what's necessary as the things that are out there that I should do.

Since I too am programmed to survive, and know more of ***ITS*** existence, I try to observe what ***IT*** is doing as ***IT*** reshapes into new events. Now that I know who is running the show I just have to be aware not to become an obstacle. If I impose my wants on how things should be rather than let things be the way ***IT*** wants I know I will get hurt, and that it will be painful to my existence. So, I try not to go up against what ***IT*** is trying to reshape into while ***IT*** permits me to be here to watch ***IT*** as ***IT*** looks to reshape into all existing possibilities.

Like I have said elsewhere, the one that started this show called evolution from ***ITS*** beginnings is ***IT***, and the only one that runs this show is ***IT***, down to the last atom that makes this show (***IT*** evolving) possible. Now I keep my main focus on ***IT***, as the way ***IT*** is reshaping, not as the things that I once thought I was doing.

Let me return to that other me that I have not yet been able to corner, for I know that it also exists at the moment that I try connecting with ***IT*** in my meditations. This other one in me is who whispers and says: "Oh you don't need to meditate, you're all right. Or it brings up a thought just as I am ready to connect to ***IT*** to distract me into thinking about whether I turned off the stove, or locked the door, or many other things.

In watching this other me I have noticed that it wants to be more in contact with what it is in my outside world. But the real me, the one who I really feel I am, wants to be with whoever ***IT*** is as ***IT*** exists within me, and I find that in being with ***IT*** as ***ITS*** oneness within me is a great feeling. When I return to ***IT*** as my outside world ***IT*** is a battlefield.

Yet, as an observer of what is happening as ***IT*** brings ***ITS*** positive and negative forces together I can see the sparks ***IT*** produces. But I no longer try to change things into what I think they should be, for I now know that no matter what I do, it is ***IT*** that has the final word as to what is going to take place out there. I know what my instructions are and I play my role, which is to do the things that are in front of me as something that I know are there only for me to do as my moment of existence.

I do not need to ask ***IT*** why I should do the things, for they are things that I am suppose to do without asking, as part of my existence, as the way I am suppose to transfer ***ITS*** energy, as the actions that I take in ***ITS*** show, for I am now aware that everything out there is ***IT*** down to the last atom, that everything out there exists as ***IT*** exists. And let me add these last words, I have also noticed that when I am inside myself with ***IT***, as meditating, ***IT*** does not talk to me, neither can I talk to ***IT***, for I would be using my mind, which would distract my focus on ***ITS*** presence. So all I can say is that in meditation all I can do is just be with ***IT*** as a place that exists within me. I have also noticed that the other one that exists as me is not allowed in this place that ***IT*** exists as; where only whoever I am can be with whatever ***IT*** is.

Analyzing the mind

The human mind does not like being analyzed. One reason may be that in analyzing the way the mind thinks, we might end up finding out that thinking is a form of nothingness.

Our **mind** does need the use of ***ITS*** weight as energy, in order to formulate a thought about what is being seen, or touched or otherwise sense, for the human mind was basically designed to see and analyze what is outside of itself as matter (***ITS*** weight)

Trivia

Contamination

We begin every year with an unspoken agreement to take part in the contamination of this planet. The smoke in the wake of the New Year's Eve fireworks is so thick that we lose our sense of the fresh night air. This smoke remains in our atmosphere far longer than it remains in our streets—it will endure throughout our lives as well as the lives of our children. However, the human race began contaminating the planet long before fireworks existed.

Our personal contribution to contamination begins about six months before we are born when our parents prepare for our arrival by buying cribs, bibs, blankets, bottles, toys, and all else that we may need as an infant. These so needed baby accessories come at the expense of land and water. In their production, they leave behind scraps and excess dyes, factory waste that must be disposed of someplace — some area of land or water.

Even when we clean, we contaminate. The chemical compounds in soaps and detergents, waxes and deodorizers, all that we use to wash our bodies, kitchens, cars, and all the material belongings we desire to keep looking new adds to the contamination of our planet. Think about all our material belongings! Our need for them leads to cutting down trees, turning islands into landfills, polluting our waters, and poisoning animals and even humans!

We contaminate for our want of televisions, radios, cell phones, furniture, and clothing. Think of the jewelry we wear on our necks to show our belief in a god or of all the religious ornaments that we hang on holidays. Multiply your belongings alone by several billion. What happens to these billions and billions of things when we are finished using them?

Is it possible for us to think positively about this contamination? Might we even be grateful for it?

Let's look at cars. The production of cars produces toxic waste and garbage in general, and the production of our one car's waste doesn't end after its purchase.

As it ages, its muffler and exhaust will also age and leak polluting fumes, so what do we do to stop it? We replace the old exhaust pipes with new ones that were manufactured in a factory and contaminated some area. Not only that, but the old parts have now become trash. So, by replacing the muffler to stop the contamination of the air in one place, we perpetuate the contamination of another area. Yet our cars are great conveniences; they get us where we want to go.

Contamination is going to continue until the day we die and continues for years after our deaths. Our decomposing bodies will leach into the surrounding soil and contaminate it. Caskets themselves are foreign objects in the earth, infusing the soil and the roots of grasses, flowers, and trees with varnishes, metals, and plastic contaminants.

You may now be asking yourself, "What can I do?"

You can be grateful to contamination; look at all the conveniences and luxuries and shiny new things we have as a result of it!

Our GOD began contaminating the universe long before we even existed. We can see meteors as contamination since they often destroy other existing areas of the universe. We can see smoke from forest fires as contamination as it robs flora and fauna of oxygen and life. We can see ash from volcanoes as contamination for dirtying the air and ground. Yet without smoke we would have no warning of fire. Without ash, we would not have the substance to make cement.

And without cement, we would not have concrete blocks or high rises or sidewalks. Contamination cannot stop, but it can be transferred. We use contamination to further our needs.

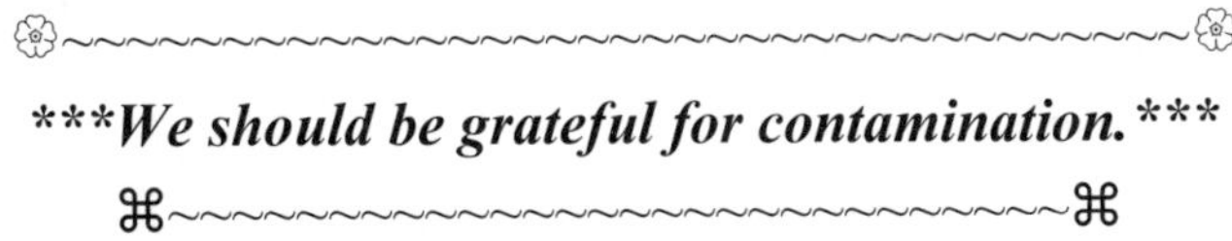

We should be grateful for contamination.

Transferring energy

It could be understood that contamination is created, that material goods are created, that even our wants and needs are created, but in this examination of the universe, it's evident that nothing is created. We do not even create music. We transfer energy from one form to another. The body transfers energy to the instrument and the instrument likewise transfers energy to the environment as notes that our ears hear and our brains perceive as music.

The instrument transfers energy in the form of notes and silence: the duality that makes music as sound exists. Einstein transferred pure energy in the form of theories. Had Einstein been born 2000 years earlier, he would have understood energy differently than he did during the moment of his existence. Einstein's theory that energy equals mass times the speed of light squared gave us the formula $E=mc^2$ and the knowledge of how to use energy in the form of work.

Einstein and his discoveries are the same pure energy that music is and that our GOD is. Einstein's lifetime, like a symphony, was a moment in the event of our GOD transferring energy.

⌘~~~~~~~~~~~~~~⌘~~~~~~~~~~~~⌘~~~~~~~~~~~~~~~~~~⌘

*** ***Environmental pollution is a side-effect of the reshaping of ITSELF as ITS weight into human beings*** ***

❀~~~~~~~~~~~~~~~~~❀ ❀~~~~~~~~~~~~~~~~~❀

UFO versus UNIFAM

Of the terms UFOs and Martians, which is more limited? I prefer the term Martians to signify aliens from other worlds, although it may still be limited. I definitely believe the term UFO is limited. If UFOs exist, the occupants maybe distant relatives of mine.

If UFOs as aliens do exist, the occupants must have a more superior knowledge and technical expertise than humans to have accomplished what they have. For our purposes, lets call these foreigners "UNIFAM," the Universal Family of Relatives. These are relatives that we have not met or do not truly understand yet. The interesting thing about humans and UNIFAM is that we each share the same pure energy as God.

It is because we share the same omnipresent GOD that we must be related; we share the same existence. Something to think about...if we were abducted, would ***IT*** be responsible? YES! As omnipresent!

Is life out there?

Is there anyone else out there in space?

I find this question limited. It is the human mind that is looking for someone with the same qualities we associate with. We want to find someone like us. This is our best attempt to understand the alien.

I find that if I focus more on who ***IT*** is, that can reshape into whatever this alien is, I can begin to understand why ***IT*** would reshape into that life form. The most important thing is not who this alien life form is but what ***IT*** is that can be all these things and still just be one.

I agree with myself, if not anyone else, that the only reason why I can question anything out there is because I, we, the mind, can only exist by virtue of the pure energy that is ***IT***.

One thing we know scientifically about life is that it can exist in a wide range of cold to hot environments and that in all life forms there is mobility as ***IT*** reshaped into that life form.

I feel that if ***IT*** can reshape into us, with a brain to think, with arms and hands to build with and legs to move with, I will not put limits as to what ***IT***, as this pure energy, can do elsewhere in the universe.

As we have gotten into smaller and smaller scales with microscopes, and now more and more nano-technology, we continue to develop different production systems, nano-factories, where we see what ***IT*** is doing in there, in such an area that is so tiny we can see ***IT*** in sub-atomic activities. It is truly amazing, the things ***IT*** does in such tiny areas, and on the other hand, as a duality, what ***IT*** does on the larger, universal scales.

As we get into a bigger and bigger scale, we can see and understand better and better that our galaxy is not even the size of a grain of sand in the vast cosmos that ***IT*** is.

IT does things in this manner. Are there any complaints?

IT is one, yet in this one ***IT*** exists in infinite arrangements of atoms.

IT takes any given arrangement of atoms and continues reshaping into infinite varieties.

And I have to say "thank you" for letting me see and understand ***YOU*** better.

And if we do find different life forms out there, we will then see ***IT*** as other life forms, as different possibilities.

IT is incredible and amazing as to what ***IT*** can do. I find that all I have to do is look and understand everything. Then I look again to determine the duality, the other side of the pure energy.

❀~~~~~~~~~~~~~~~~~~~~~~~❀❀~~~~~~~~~~~~~~~~~~~~❀

***** Oh God help me to understand you more, and never let me resist your reshaping, even if it produces hard times in my existence *****

⌘~~~~~~~~~~~~~~⌘~~~~~~~~~~~~~~~~⌘~~~~~~~~~~~~~~~⌘

Our minds and extraterrestrials

Here are some ideas that relate to our never-ending question concerning whether or not there are extraterrestrial beings out there in space.

First, let me remind the reader that no matter who exists out there as extraterrestrials, they have to be, to those that believe in God, our relatives, since they too are made from ***ITS*** weight and from ***ITS*** empty nothingness. How else could they exist? If they exist, they have to be made of matter (atoms).

But here is the main point when it relates to there being extraterrestrials out there, to which I have to remind you that they are not "out" there, for they too have to exist *within* ***ITS*** nothingness. Empty nothingness that now exists as this empty universe, as omnipresent

Our minds have been so accustomed to the way things are here on Earth that we have organized programs and made films where we are at war with others (***ITS*** weight) except in some cases such as the film called E.T. We can still see that our minds think in terms of war-as in Star Wars, Star Trek, War of the Worlds, and other movies. Our minds have not readily accepted that we can exist in peace, as in meditation (***ITS*** nothingness), and as the peaceful groups we have on this planet, like priests, monks, and many others.

We can see that our minds are curious as to the existence of extraterrestrial life forms, about which I have to remind the reader that anything that can exist as life is really just ***IT,*** even though we have been trained from our earliest moments to see extraterrestrials as enemies. And I must say that up until now we may be the only life form that will be traveling within ***ITS*** other parts as its interior (outer space); at least until some extraterrestrials send us a message.

I hope that they are aware that they exist as ***IT***, as pure energy.

We should try to remember that we have the know-how to be able to exist in a state in which our minds are not in a constant battle, so it might be possible that as we leave this planet we can do it without having to adjust our minds for a battle with someone else. But we will have to continue with our two basic programs: survival and reproduction. ***IT*** will find what will be necessary for our new bodies to reshape into. The same way we do not look like our earliest ancestors, our space bodies will not be the same as our planetary bodies, and if we stay in outer space too long, when we return, we might even be mistaken for extraterrestrials.

And yes, it won't be boring as we travel in outer space (***ITS*** interior) without wars with other life forms, for there will be other problems that our minds will have to attend to in order to fulfill our 2 basic necessities.

Evolution

Have you ever noticed that when you are not in search of a car or an apartment you rarely notice postings for them? When you are not looking, you are less likely to find anything. The same applies to evolution. When we experience life and nature without looking for it, without paying attention to anything in particular, we do not notice the changes involved in our own evolution.

Evolution is a transferring of energy. ***IT*** is the reshaping or transmutation of things around us. For instance, birds with short beaks cannot eat from fruiting plants with large leaves. Over time, those birds with longer beaks will be the ones who most successfully survive to reproduce because they are able to reach the fruit. This is what scientists call "natural selection". If those birds with shorter beaks don't find an alternate food source they will become extinct. Likewise, those plants that consistently produce shorter leaves will survive because the birds will be able to eat their fruit, thereby spreading plant's seeds.

❁~~~~~~❁

God does not emulate nature, for God and nature are the same.

⌘~~~~~~~~~~~~~~~~~~~~~~~⌘~ ~~~~~~~~~~~~~~~~~~~~~~~⌘

IT will even make ***ITSELF*** bisexual where there is a lack of one or the other gender in order to continue propagation and the survival of a species as seen in anemone fish that live in small groups in which only the largest female and male reproduce. If the female dies, her mate becomes a female. The largest fish in the group, which by the way, hasn't selected a gender yet, will become a male and mate with this newly transgendered female.

ITS transformation of the tree's leaves, the bird's beak, and the anemone fish's reproduction are perfect examples of evolution. When the plant and bird or fish are no longer necessary, ***IT*** will reshape their energies into something completely different. We see this as the destruction of a species, but this is natural law according to ***IT***.

This law states: "Anything that is created must be destroyed." Without destruction, one thing cannot reshape into another as is needed as that moment. I discuss this in more depth in Sections Two and Three.

❁~~~~~~~~~~~~~~~~~~~~~~~❁❁~~~~~~~~~~~~~~~~~~~~~~~❁

** ***I now exist where things are not the way I want them to be, but rather, as they are in ITS reshaping.*** *

⌘~~~~~~~~~~~~~~~⌘~~~~~~~~~~~~~~~~~⌘~~~~~~~~~~~~~~~⌘

Law

Science tells us there are different laws governing our existence. One is the Law of the Universe: Anything created will eventually be destroyed. Another is the Law of Change. This law follows the trend that the only thing that does not change is change itself. Change is pure energy reshaping ***ITSELF***; ***IT*** continually transmutes into something different. Consider survival and reproduction, the two programs conducting the existence of human beings, and how they have changed.

Humankind's earliest method of survival was hunting and gathering. After we became more organized and communal, we developed agriculture. Survival was all about food, and food was all about survival. Once these methods of sustainability were established, humankind then produced teachers to distribute our knowledge as well as engineer an infrastructure for advancing that knowledge in the future. The Egyptian pyramids for example showed us our first waterproof ceilings, we can see how advanced knowledge then led to the greater protection and height of the modern, flat roof. Every time energy is transformed it courses through the receiver to make a change.

Another law is the Law of Nature, now officially known as Isaac Newton's Third Law of Motion, which tells us that every action has an equal and opposite reaction. These actions and reactions could also be perceived as the duality of positive and negative forces that started before the Big Bang.

IT transmutes the laws of nature into the laws of man. This transmuting or reshaping is the pure energy of GOD at work. The laws of man are in place to ensure human survival into what we think of as the future. Because we see things in ***ITS*** own image, we see this same positive and negative force in our systems governing humanity. We see these forces at play in the duality of prosecutors and defenders in the system of law and justice.

❁~~~~~~~~~~~~~~~~~~~~~~~~❁ ~~~~~~~~~~~~~~~~~~~~~~~~❁

******We all have a need for change. Certainly, you have heard people say that they need to make some changes in their lives. These changes exemplify how we are made in ITS own image because IT is constantly reshaping ITSELF.******

⌘~~~~~~~~~~~~~~⌘~~~~~~~~~~~~~~~~⌘~~~~~~~~~~~~~~~~⌘

Law as a profession

Our education system has produced an abundance of teachers and engineers, and now ***IT*** is providing an excessive number of lawyers. And similar to any supply that's a dime a dozen, consumers need to be extra alert of quality. Consumers would be well off avoiding some lawyers, or prepare to spend their hard earned and rainy day dollars on them. Lawyers can be like a pair of scissors, especially in divorce court. Regardless of who wins, the clients in the middle of opposing lawyers get cut. Former partners become enemies. When initially choosing our partner in marriage, we don't consider we could also be choosing our future enemy! The arena of political law exhibits another interesting duality of forces, for wherever one is there are extremists on either side: an extreme right and an extreme left. It could thus be said that we are all somewhere in the gray, in between the two extremes.

Should a crossing with a wayward lawyer happen, we must be prepared to spend our hard-earned and rainy day dollars, as we may end up at their favorite meeting place, the courthouse. Be especially cautious of some lawyers who have recently graduated. They are looking to take any case that will give them a chance to practice in front of a judge.

Two main reasons so many men and women enter the field of law is to become wealthy and to gain recognition as important persons. With one high profile or winning case, a lawyer could become a millionaire. Accordingly, the party represented by the ambitious lawyer, likewise files suit for the purpose of wealth and recognition. One thought that I have after watching lawyers in action is that speculation does not allow for justice.

Fortunately, or not, ninety percent of the people in law will not become millionaires at the expense of political challenges or divorces. They will become entangled in the bureaucracy of law processes and end up working in a different field of public service altogether. Since all lawyers have advanced formal education, they have the ability to use it in other fields--government agencies or private industry.

Justice or revenge?

The concept of justice, like everything else in this Universe, has evolved over time. A certain website defines justice as something which is constantly strived for and rarely achieved.[3] It is a fact that humans are a selfish lot, at worst thinking only of themselves and their own needs, and at best extending that concern to their own

---

[3] http://jove.prohosting.com/mshambli/UniversalADG/j/j.htm

immediate family. There is no doubt that we still retain the territorial animal instinct.

In order to curb our selfishness and territorialism and to enable us to live together in relative peace, laws were created. A system of law determines rights and assigns punishments for their violation. Rights are based on humanity's two basic programs: survival and reproduction, the fulfillment of which requires access to the environment's natural resources. Laws are enacted to insure that each person gets his or her fair share of these natural resources.

Our concepts of justice are based on this feeling of "fairness". In fact, this internal sense of fairness was once used as an argument to prove the existence of God.[4]

When this state of "fairness" is upset, people feel the situation must be redressed. Now, in Antiquity, before the rule of law, this redress took the form of vengeance or revenge. In those days there were no limits to revenge, and retaliation for wrongs received or perceived could and did very easily take on a genocidal character. Even in more modern days, and in our supposedly civilized and enlightened society, during the late 1800's, the Hatfields and the McCoys of Tug Valley in Kentucky had a long lasting feud in which they took turns killing members of each other's family. Would you believe it all started when Randolph McCoy accused Floyd Hatfield of stealing his hog? It was to stop this kind of thing that the old Sumerian dictums "an eye for an eye", and "a tooth for a tooth" were enacted, later finding their way into the Hebrew Scriptures.[5] These laws were a good thing in their time because they limited vengeance. A person could not exact vengeance for anything more than the injury received.

The next step forward was the creation of a judicial system: impartial judges and courts, where grievances were aired. Vengeance was taken out of the hands of individuals and put into the hands of a supposedly impartial institution, empowered by the state to administer the applicable laws.

---

[4] Immanuel Kant's Moral Argument

[5] Babylonian Emperor Hammurabi ca. 1780 B.C.E.; who codified earlier Sumerian laws.

As justice evolved, so did the concept of who is to be considered a person. Not everyone had the right to appear before a judge or a court of law. Back in the old days, slaves were not persons; sometimes women were not persons at all and at other times they only achieved the status of second class persons by virtue of who their husbands or fathers were. Children were not considered persons, either, and under some systems of law they were in the same class as slaves. Foreigners were not much better off. Besides this, among those who were considered persons, not all were considered equal. Laws were different according to the social status of the persons involved, and therefore, penalties or punishments for breaking the law also varied accordingly.

Breaches of the law are carefully classified into criminal offenses (felonies and misdemeanors) and civil offenses. Civil matters involving negligence are usually settled by the applying penalties for damages that are paid for with monetary energy.

This is the way of human justice. When judges and courts apply merely the letter of the law they may commit injustice. This is where equity or mercy comes in. In some countries, mercy is considered a weakness and wrongdoers are punished severely in order to set an example to others. In short, this is law and order based on fear.

Can’t there be a better way? As Gandhi said, "An eye for an eye and a tooth for a tooth leaves the world blind and toothless." It is nice to see that masters like Christ knew the dangers of revenge, for this revenge will contaminate the person that wants revenge. Desire for revenge is based on feelings of anger and hate; on wanting to do unto others what they have done unto you and then some. Forgiveness, however, is based on love; on the admission that no one is without fault. But just as a hole remains when you remove a nail that has been driven into a piece of wood, forgiveness does not take away the consequences of wrongdoing. One must make amends.

Inmates’ rights are another sticky issue. Clearly those who have been rightfully convicted have infringed the rights of others, so it is only fair that their rights be taken away, that they may understand and learn that one’s rights end where one’s neighbor’s rights begin.

But how many rights are to be taken away? Are there not certain inalienable human rights, which when taken away leave a person in a less than human estate? The right to food, clothing, water, bathing, exercise, sleep, to not undergo torture…But what of those whose crimes have involved precisely depriving others of the enjoyment of these very same rights?

Based on the analogy of parents socializing children, our modern day concept of a prison involves re-educating, reforming, rehabilitating those who have transgressed the laws of society. This is another one of those things which is constantly strived for and rarely achieved. Wrongdoers have proven themselves unfit to live in society, so they are taken out of it. They are either removed to a penal institution where rehabilitation is attempted, or they may be considered beyond all hope, and are given the death penalty, for the dead are no longer a threat to society. The downside of life in prison is that it is we, the people, the law abiding citizens who, through our tax dollars, are required to feed, clothe, and house the offenders… And those who oppose the death penalty don't realize that they are thereby indirectly approving the death of the victims.

Truly, human justice is a very messy thing.

The power of the minority

In a typical democratic society that is more or less split down the middle on most issues, a small minority can have the power to force change.

A political party that controls anywhere from three or four percent to nine or ten percent of a voting public can hold the bigger parties hostage to their demands. At a minimum, they can make the bigger parties court them.

We can see this kind of sway power held by minority parties in countries that hold democratic elections. We read about coalitions that organize under one issue in Italy, France, and Belgium. In fact so many countries have maybe eight or ten parties striving for power in the government, it's a wonder how countries accomplish anything!

I realize that it is difficult for some people to understand me when I state that I am politically free. I am neither a capitalist nor a socialist. To be part of a group is to exclude yourself from the total population. Once you are willing to accept GOD as the total, you no longer need groups.

Everything is the same. Everything is pure energy in different forms. The energy inside and outside of me makes everything possible.

Overpopulation

While the world's population could fit in the state of Texas, it couldn't sustain itself on Texas' resources alone. It becomes clearer and clearer that most of the damage to the world's ecology results from the expansion of the human population into the natural habitats of the rest of Earth's creatures. Every day we can read articles or see documentaries showing how another species of animal (large and small) is on the verge of extinction due to people moving and living in the animal's habitat. This is happening, as I write and you read, in Africa, India, the Amazon and other parts of the world, including the Caribbean and the USA.

The need for more timber for more building leads to the destruction of more forests all over the world. The need for more fossil fuels has humankind drilling into more and more sensitive environments all the time.

All of this destruction is due to the demands of an ever-growing population.

The human race adapts to overpopulation by creating more of the material objects that it needs to survive. This trend will not change. With every generation, energy transmutes ***ITSELF*** to adapt. Likewise, ***IT*** has also established population controls: the Great Flood, the Black Plague, AIDS, and war along with other devastating illnesses and events.

The "population explosion" as it's been named is part of the perfect plan. Humans are programmed to survive and reproduce. Men would be hard pressed to convince women to stop having children.

Even liberated women want to experience the miracle of childbirth; even liberated men want sons to carry on their names. And technology has become so advanced with artificial insemination as well as artificial wombs that we are even able to produce children without parents.

Nature guarantees human reproduction (as much as nature can guarantee anything). Humans, male and female, are attracted to each other because they are opposite in polarity. When drawn together, nature assures a union; all we have to do is follow the heat.

The next time you hear the media announce that the poverty level has gone up by five percent, be aware that this means the population has gone up five percent. The more our population goes up, the harder it becomes to feed it. It will become even harder for people without incomes.

⌘~~~~~~~~~~~~~~~~~⌘⌘~~~~~~~~~~~~~~~~~⌘

*****GOD is running this show; we are only pawns in ITS plan*****

❀~~~~~~~~~~~~~~❀❀~~~~~~~~~~~~~~❀❀~~~~~~~~~~~~~~❀

IT as overpopulation as a maximum

We now feel the stress generated by overpopulation: our lives, our environment, streets, cars, trains, buses; the destruction of planet resources, insufficient housing, lack of food, and more and more, not enough jobs.

We see the contamination of our planet, caused by more than five billion people.

It is estimated that in two decades the world's population will double, and since there are two women to every one man the population could more than double.

I know that ***IT*** knows what ***IT*** is doing in taking our present environment as a society from a minimum to a maximum. We are forced to adjust because we are genetically programmed to reproduce.

In the USA Social Security was invented to provide a means to subsist after we stopped working. But this was based on a theory. The theory was that there would always be more people coming into the workforce that would support the retired population. Now a days we see that many have to go back to work because Social Security does not cover present needs, and even more to the point, less people are coming into the workforce in relation to the total population; so less people are supporting the payment system.

Industrialists control industry and jobs. A good example is how a great majority of products are produced with planned obsolescence. Technology today makes it possible to produce items of much better duration, yet many products are still made to be discarded very quickly. The point is to keep us needing to buy more (many times the same) products in the future.

As we know, we live in a throwaway economy. It is hardly worth repairing many of the articles that used to be repaired. Household appliances, computers, printers; most of these things are throwaway. As a consequence, it would be better if governments would push for higher minimum wages; that way they could collect more taxes from the few who have been able to keep their jobs.

Even many that have been able to keep their jobs have found it necessary to dip into savings (what I call stored monetary energy) in order to survive.

Having a job, we have had to use some of our stored up monetary energy in order to continue surviving. ***IT*** knows what ***IT*** is doing, and no matter what happens all we have to do is stay as close to ***IT*** as possible, and always ask for guidance in what to do as ***IT*** reshapes.

There have been other disastrous moments in our history. I do not feel that we are close to the end of humanity, but rather just like in other times in history, we will make the necessary changes to be able to become the new, high-tech society that will survive into the future.

In the agrarian stage of our existence it was said that "where one can eat, ten can also eat," but now things are not the same. We know that not even two people can eat well, let alone four or six. And to be honest, it does not look like it will be getting any better, for the cost of food and shelter only gets higher.

I sometimes think of the wars that are going on right now. From what I can see they are waged in the hope of finding more oil wells that will fuel our autos and warm our homes, make it possible for jets to take us on vacations to other lands…

Women & men

If there is a group that can save the world's economy and environment, **it** is women. Women can help put a stop or at least a limit on this destruction by taking control of their bodies! Stop having so many children! Don't let men impose their will and bodies on you--ask them to have the operation. Defy the programming you have to reproduce!

Generally women want two children: a male and a female. Some men have more than one parenting partner too: a wife and an ex-wife for example, each typically with or wanting two children. Just as women must take responsibility for contamination in the home, men need to take some responsibility here since contamination begins with our own population. A man who marries twice could curb overpopulation and excess mouths of consumption by having a vasectomy.

The procedure only takes 15 minutes. I know because I had it done when my first wife wanted ten children. I told her that if I could not learn from two children, I certainly would not learn from ten. Furthermore, a government conscious of this important issue could help offset the cost.

Women were the first to picket the White House. They may be the answer to our population crisis and numerable other world problems.

Women make such an impression on people that the figure of a woman has been honored with representing justice on a global scale. But Justice is blindfolded! While intended to show the fairness of law, here we are reminded of our self-limiting perceptions. We must take off the blindfold in order for the dualities to exist. There is no true democracy without woman wielding the same power as men.

What about a female dictatorship? Here is how it could happen: Men go to war and die in large numbers. Even without war, men do not live as long as women do. There are currently more women than men. If women would unite, they could dictate to the world. There is no better moment than now. It’s the perfect time, in fact, for women to take over the political infrastructure. Women now have the opportunity to take over the current political

infrastructure. To do so, they need to unite, organize, and collectively take advantage of the civil liberties they worked so hard to gain.
The fact that men dominate the government currently is not because women are not welcome; it is because women, as a group, have not used being a majority as a voting power to make a difference.

⌘~~~~~~~~~~~~~~⌘~~~~~~~~~~ ~~~~~~⌘~~~~~~~~~~~~⌘

It bears repeating: IT is running the show; we are only pawns in the overall plan

❁~~~~~~~~~~~~~~~❁~~~~~~~~~~~~~❁~~~~~~~~~~~~~❁

On that last note, could it be possible that because of the personal differences in the perception of men and women, computers are more likely to take over the political infrastructure? Computers are not bigoted in their thoughts, be they racial, sexual, political or anything else.

Thinking more about women and men, since women out-populate men by sixty percent, is there not a good possibility that we would see more women in areas where they have not been seen? In theory, a woman could win the next USA presidential election by a margin of six to four. And if homosexuals gave their vote to women, a women president would arrive sooner. Whose blindfold is still on?

The door has opened, perhaps it is time for women to charge in and clean up the mess made by men in the government the way they have had to clean up man's mess in the home.

Did you know that Adolph Hitler did more for the liberation of American women than any other single person did? It's true. Consider this: All able-bodied men from the USA went to war while the women remained at home to run the industries that were previously male run. When many of the men did not return, women continued to staff those jobs. What's more, women began pursuing other careers previously dominated by men.

Hitler and the black man

Hitler, unknowingly, gave the black American male a push in the direction of getting more equality at home. Having been recruited to fight in WW II, black men fought equally next to white men when they returned home the struggle for equality continued.

❁~~~~~~~~~~~~~~~~~~~~~~~~~❁ ~~~~~~~~~~~~~~~~~~~~~~~❁

****Adolph Hitler impersonated Charlie Chaplin's moustache.****

⌘~~~~~~~~~~~~~~~⌘~~~~~~~~~~~~~~~~~⌘~~~~~~~~~~~~~~~~⌘

Robots and the equality of women

In order to understand this better, we should start with men and their dominance in some sectors of the labor force. At the beginning, men had a stronger hold on jobs that where demanding of brute physical strength, but now robots have been brought in to perform tasks where brute physical strength is a job requirement, as in lifting or moving of very heavy materials, which men could physically perform better than women. Things have changed, and now robots can be operated by women. It is logical to think that as more and more women operate robots, they will both do the job better then men did it before. The next step will be that women combined with robots will be replacing these men; because our social working structure (the capitalist system) always tries to find someone that will do any job better and cheaper, which is precisely what is now happening.

Now I would like to add that what is happening with in our social working structure, with women having to work more for less money, is something that has been here as far back as the Industrial Revolution. What is happening to women today is the same thing that industries where doing to men back then. This is a situation that still exists where there are jobs that have to do with men only, for this system will exist as long as there is someone investing his/her money, where whoever is investing wants as much work done for the least possible amount of money.

I think that women should remember that what is happening to them now, also happened to men; to men among men who had to endure the experience of other men that forced them to accept a lesser pay and possibly more hours than other men who were being paid for the same job.

And now that women are able to operate robots that are doing the work that had to be done by male physical strength, they find it easier to be equal with men in the labor force. In fact, let me add that as women are taking work away from men, men are also doing jobs that were mostly done by women, such as typing, for we can see that there are a lot of men typing away at keyboards nowadays.

Cloning

If today there existed a clone of Adolph Hitler, it would not be the same Adolph Hitler of the past because this is a different moment as time. Since the time of Adolph Hitler, the world has reshaped itself***;*** therefore, he could not repeat the past.

Men are in real trouble.

Women complain about men. They complain that men lie. It is true; few men are saints, but are there not similarly few sainted women?

At home, women become upset when men leave the toilet seat up. Putting the seat down shows consideration and respect; something that many men lack towards women. Nevertheless, eighty percent of men leave the toilet seat down.

The toilet seat has become a major battle for men. Before it was invented, did women or men complain about not having it? I believe the invention of the toilet seat created this problem for women and men. To end the battle perhaps we must return to outhouses.

I believe the world is a warmer place because of women.

But why do we have a Miss and Mister Universe pageant? Shouldn't it be Miss and Mister Earth? After all, there are no aliens in the contest. (Unless we simply cannot see them!)

Women want to have children. As humans, men have to impregnate women so that they can give birth, and women must be receptive of men in order to conceive. Ninety percent of women are programmed to participate in the second human priority, reproduction. Giving birth allows pure energy to continue ***ITS*** reshaping. It also gives women and men something that they can call their own.

Adults lose a great quality as they grow up. Children have values. Children in general are simple minded and forgiving.

Children do not become involved in politics; they enjoy the simple things in life. We say that children are little "angels." What happens in life to turn them into the little "devils" that adults become?

❁~~~~~~~~~~~~~~~~❁~~~~~~~~~~~~~~~~~~❁

***** Some things will resist change, such as the toilet seat.*****

⌘~~~~~~~~~~~~~~⌘~~~~~~~~~~~~~~~~⌘~~~~~~~~~~~~~~~⌘

But is this to say women and men are equal?

I believe not in the sense that women and men are not the same—not physically or mentally.

Women are more accepting. Men are more competitive.

Women tend to trust the public majority; men do not.

Women do things eighty percent differently than men. Consider this: a man might go to a party and without hesitation try to sleep with a woman old enough to be his mother. If it were his mother, it would be a different story. The same applies if he approaches a young girl. If it were his daughter, it would be unconscionable. This shows us something about the dual mentality of the mind.

The equality of men and women is in the balance; ***IT*** is in their duality. And they love and battle each other to maintain the balance. Women have worked for years opening the door, and it's now ajar. It's time they charged in and cleaned up the mess made by the men in the government the way they have had to clean up man's mess in the home. On that last note, could it be possible that because of the personal differences in the perception of men and women, computers are more likely to take over the political infrastructure? Computers are not bigoted in their thoughts, be they racial, sexual, political or anything else.

❁~~~~~~~~~~~~~~~~~~~~~~~~❁ ~~~~~~~~~~~~~~~~~~~~~~❁

******Most adults actually believe they belong to their parents. Most do not believe that they belong to God.******

⌘~~~~~~~~~~~~~~⌘~~~~~~~~~ ~~~~~⌘~~~~~~~~~~~~~~~~⌘

Computers

Computers are widely accepted now. Computers are running most of the world's infrastructure and will eventually run the world's entire infrastructure. They are not human; they have no feelings. Computers have no political preference or interest in personal relationships. They do not get weary, lazy, need a workers union or benefits. Computers do shut down; they do have flaws. They are pure energy as ***IT*** reshaped. Computers are also made of atoms. These atoms that are the computer exist in this place called omnipresent. Computers transfer energy.

Electronics

We are more electronic than any equipment that exists. We are made of more atoms than most computers. The word electronic stems from the word electron. An electron is the outer-most part of an atom. We have all that in our bodies.

Money

Money is monetary energy and work is human energy.

The majority of us work because of the necessity to support ourselves financially.

Consequently, many people put in long hours doing something they do not enjoy. These hours of busyness are an expenditure of energy, human energy. This human energy could be spent either mentally, physically or both.

This human energy is what most call work. Money is a piece of paper, or coin, or plastic that carries an invisible energy, activated through our belief in it, our holding of it, and our spending of it. This monetary energy has become necessary for our survival.

Once you have monetary energy, you can transfer it as you please. One of the primary things we exchange it for is food, which provides us the energy that we need to maintain our survival. As earlier mentioned, our first programmed need is survival and our second is reproduction. We survive so that we can reproduce, work, earn money, and spend it; it is a cycle of transferring energy.

We are created in ***ITS*** own image; we are part of the cycle of transferring energy.

Another way to look at money is like this: since money is energy, it can perform a kind of work. That means that one penny has as much energy attached to it as the amount of work needed to be performed to earn that penny.

Our first and second basic necessities, oxygen and water, are free because GOD provides them to us. From the beginning of human development water was free, but we had to find it. Water is still free; it comes in the form of rain. However, there is a price attached to the receipt of water--a price dependant on the amount of energy required to receive the water in your home. GOD also provides food, but it too requires the use of human energy to ensure that its location is close enough to where we live.

In physics, energy is defined as something that can do work, and integral to work is "necessity," hence the coined phrase, "Necessity is the mother of invention." I say that necessity makes us do things that we would not do other wise. Necessity also ensures the transfer of energy.

If you do not like what you are doing, ask your God, "What else YOU can do?" Additionally, you must let go of what you are currently doing and let GOD show you what you need to do.

You may wonder about the portion of your monetary energy that you pay to the local and federal government in the form of taxes. This too is part of ***ITS*** reshaping, and for this too we should be grateful.

Excess money

Obviously, people with an abundance of money are able to do more things than those that barely get by. Affluent people, for example, have the ability to have exploratory surgery if recommended, while other people cannot afford such options.

Could this excess money be a trap? Could it be that because they are well to do, they can do more? Could it make them less likely to cherish the most important things in life, such as health and family?

One thing is for sure; the affluent are permitted to do what the majority of us cannot do in the participation of ***ITS*** reshaping.

Think about the saying, "Time is Money." This is a recent human concept. Money did not exist for primitive humans.

We all want to make the most amount of money that we can. And, we want to spend the least amount possible when buying material things.

Competition in the market of materials controls the price for the consumer. Competition in the market controls the amount an employer is willing to pay an employee. Salaries in general are higher in areas of high employment than they are in areas of low employment.

In the USA a person working as a courier will make more money for doing the same job than a person in Haiti.

The global economy has changed the face of what is acceptable pay. The dollar is not worth what it was 50 years ago. Inflation affects everyone. Due to inflation, the money that you have saved for your future will not likely be what you need to survive at retirement.

The hard work that earned you $3.50 an hour has less value today, for the item that cost you $.99 cents when you were earning $3.50 per hour now costs you$1.40.

The effects of the recent rise from $3.50 to $5.35 per hour minimum wage in the USA is not good news for the people that have saved money. If you have saved $1000.00 from the time you were making $3.50 per hour, your $1000.00 is now worth 40% less in buying power. Again, what used to cost $.99 per pound now costs $1.40 per pound.

It is all about supply and demand. The more we make, the more things cost. When the salaries in industry go up, the cost of the product the industry supplies also goes up.

Let's look at salary as it relates to a home purchase. If you purchased your home for $50,000 when you made $100.00 per week, and your home is now paid. If your salary has since increased to $200.00 per week, the value of your home has increased to $100,000.

You have earned equity on your home, something you would not have done if you had chosen to rent. This is great if you choose to stay in your home. You may at sometime decide to move because your house is too large or your children have grown and are now on their own.

You think, "Well I have $50,000 equity in my home; I can sell it and use that money to buy a smaller house." The smaller home that you want now costs $100,000. The new home was built on more expensive labor. So, in essence you didn't profit at all.

Who do you think will profit on your house? Could it be your children? They will believe that you left them everything. They should learn, or hopefully they will learn, that when they part this Earth, the only thing that has been here since the beginning is this pure energy.

The physical property left behind continues to exist in the moment. It will stay within the omnipresent so that it does not violate the Law of Conservation of pure energy. This is the reason why we can never take anything with us when we die, not even our body.

When you compare stores for things like imported VCRs, TVs, and cars, to get the most out of your money, you are supporting jobs of those in foreign manufacturing and taking it away from your local economy. We should be grateful for this. This is something that we cannot control. ***IT***, as pure energy, is making this happen during a stage of ***ITS*** reshaping.

As for the minimum wage, when wages go up we get more unemployment. That means that the fewer people that have work will have to pay more from their salaries as taxes. Which is better, more people working for less or less people working for more? Which is better for government coffers?

Have you ever noticed the amount of people that are walking the streets asking for help in the form of monetary energy? All these people need to do is ask for help from the pure energy that is inside of them. ***IT*** is there just waiting for their call for help. If they call, the help will be provided to them, and let us be grateful that you and I are not in that situation, for ***IT*** can have us there also at any moment of ***ITS*** reshaping.

❁~~~~~~~~~~~~~~~~~~~~❁❁~~~~~~~~~~~~~~~~~~~~~~❁

******Some people are satisfied to call a cardboard box a home. For some, a home is the place that has walls, a roof, and furniture. Some people live in extremely large homes and never use some of the rooms. We should be grateful that IT provided our first home with a roof that permitted us our fireplaces that kept us warm and enabled our safety.*** ****

⌘~~~~~~~~~~~~~~~⌘~~~~~~~~~~~~~~~~⌘~~~~~~~~~~~~~~~⌘

Jobs

Henry Ford is the one that made many of us lose our jobs, because it was he who put machines into full speed, for he once said that machines should do the intolerable jobs that man has to do., so much so, that today it is machines that are doing many of the things that are intolerable to workers. Thanks to ***IT*** for bringing Henry Ford into existence, for machines have shortened production time, because now it takes less time to produce than before machines came along, again so much so that we now have more time for play, but without pay…

The business game

The business game, or at least from my point of view, as ***IT,*** what we call "business" is just one more game that some of us can play. Some players get so involved that they see this game as a matter of life and death, and others, like me, remember that this temporary game has a name as its owner, which we call GOD. When we get lost in playing this game, all we have to do is remember to go back to ***WHO*** established this game called life and remember that it is just a game, where we can play, but not to take this game very seriously.

For as you, the reader, will see later on, in section #2 which explains this better, you will be able to understand that what is really happening is that we are just moving ***ITS*** weight around, so that ***IT*** can continue to reshape. We should also remember that there was a moment where we as humans did not yet know of such a game, for we were still together as a group, inside a cave. In the same way, this game might cease to exist as ***IT*** continues ***ITS*** reshaping, and maybe, when we leave this planet, this game might not continue, for we should remember that our minds get into the habit of seeing things as if they have always been a certain way.

So, the same way the business game came into existence, it too might come to an end, when ***IT*** decides to reshape into where ***IT*** will no longer need the effects that this game we call business produces.

Wants

Nothing on this planet belongs to us. The day will come when you are dying and you realize that you cannot take the material things with you.

We work hard to get what we want. If you want to work less, all you have to do is diminish your wants.

If you have a million dollars, does it help you sleep better? Does a million dollars make you a happier person?

You do not have to have a million dollars to enjoy the life of a millionaire; all it takes is being able to give without question of whether it will cripple you. Furthermore, if you reduce your wants, you will not need as much money as you think you do.

⌘~~~~~~~~~~~~~~⌘⌘~~~~~~~~~~~~~⌘~~~~~~~~~~~~⌘

******The duality of our human mind causes us to want to get as much as we can for our services, but we do not want to pay as much for other people's services as their work ******

❀~~~~~~~~~~~~~~~❀❀~~~~~~~~~~~~~❀~~~~~~~~~~~~~❀

My life as a boat.

I feel that this is a way to express my feelings, when it relates to ***IT*** and I. Let me explain, why a boat, first **it** is ***ITS*** boat to begin with, and I have the freedom to decide if I want to continue in this so called normal life, normal because the majority are there.

Or as I have, surrendered to ***IT***, and in order to do this, you have to forget about what you want, so in order to sit in ***ITS*** boat you will not do the steering. First, because you will no longer know where you are going to next, as your encounter with other humans, and second, this boat is made for you only, and the boat will not carry a lot of unnecessary items, that we accumulate as we travel, on this road as our existence. And if we try to bring this extra unnecessary baggage, we will sink sooner, or not move at the speed that ***IT*** wants to move at. We have to trust in ***IT***, for ***IT*** is ***ITS*** boat and water, and even the wind that will take us to where ***IT*** wants us to be, for ***IT*** is in ***ITS*** own water.

The uncertainty road

As most everyone wants to feel secure and so many of us are born into a majority that seeks security outside ourselves, so many of us never learn there is an alternative: Trusting ***IT***. Instead of trying to know about every thing that is happening around us or trying to predict what will happen next, we can instead quiet the mind and take the uncertainty Road.

On the Uncertainty Road, everything is new. The only danger that exists is permitting the mind to tell ***IT*** how things should be and where ***IT*** should go. But it's our choice.

Eyes

Our eyes cannot see; it is the brain, as an organ mostly made of water, that actually sees. The eye is also made mostly of water. If you were to drop water on paper, under the right conditions, you would notice that the water acts as a magnifying glass. Glass is also made with water. The mind, using this magnification process forms a picture in the back of the brain.

This process of seeing becomes possible with the help from the hydrogen atoms that exist in the water, which are formed in outer space. The magnification created by our eyes allows us to see what occupies space outside ourselves. Thus, it is not the eyes that see; the mind sees because of this magnification.

This would mean that the mind had to have come first.

The white part of the human eye serves as a reflector. Anything white reflects light. The opposite of white is black. Black absorbs light. These things are necessary for the mind to see.

We cannot create

Albert Einstein did not create anything; he simply understood this *pure energy* differently. Einstein understood how to use this energy in the form of work. He did not understand why he worked so much; he just worked his mind to the maximum. His lifetime was the time for us to understand ***ITS*** energy as it refers to Einstein's work.

Einstein's theory, that energy equals mass times the speed of light squared, gave us the formula $E=mc^2$. If Einstein had been born 2,000 years earlier he would not have been able to see and understand energy as he did during his moment in existence.

Because of the discoveries of Einstein, we have more information on energy and matter. We have a better understanding of this pure energy and a better understanding of that which I call our Creator.

How about this: We cannot create music or anything else; what we do is transfer energy.

We should remember that we are ***IT***, as this pure energy that exists in a place called omnipresent. We transfer energy from our bodies to the instrument that we play. The instrument transfers energy in the form of music. Music, as a sound, is possible because of ***ITS*** duality. Music is only possible because of the silence that exists between the notes. The orchestra and the instruments are examples of the Creator transferring ***ITS*** omnipresent energy.

⌘~~~~~~~~~~~~~~⌘⌘~~~~~~~~~~~~~~⌘~~~~~~~~~~~~~⌘

*** ***God does not create, IT is every thing that exists .****

❀~~~~~~~~~~~~~~❀~~~~~~~~~~~~~~❀~~~~~~~~~~~~~~❀

A friend that is not needed

Some friends we do not need as friends. Some people only call themselves friends as revealed when the time comes that we need a great deal of help from such friends. Let's say you lose your job and you schedule a garage sale to make some extra money.

Some of your friends come to the garage sale. You have a radio for sale for $20.00; your friend says that since he is your friend you should sell it to him for $10.00. You would think that if he were a real friend he would offer you $30.00 so that he could help you out.

Life as a game

Let me explain why we refer to life at times as being a game. But first let's go over the meaning of the word game. As any game, we know that at times we will win and that at times we will have to lose. That is why they are called games. When we apply this to our existence we sometimes feel that life is a game. One reason for this is obvious, we can see that at times we gain something, and that at other times we lose something.

For I too saw life this way, until I came in direct contact with ***IT*** and how ***IT*** exists.

Let me explain it this way, before I too saw the things that exist outside of me as objects and as possessions (property), and that the more I had of these possessions the better I was suppose to be. To which I should add that this is partly true when it relates to our basic necessities. And to this I also have to add that life as this game that is going on, as in taking place, is something that does exist, and will most likely continue to exist because it is a very powerful force that ***IT*** uses to make things happen so that ***IT*** can continue reshaping ***ITSELF*** into something else, as ***ITSELF***. I know that this is a little hard to accept, but we should remember that everything in this whole universe is ***IT***.

There was a moment where we did not even exist here on this planet, and that it is because of ***ITS*** existence that we came into being where we now can play this game that exists out there which is governed by the ingredient that ***IT*** placed in us called necessity, which is one of the factors that makes our society, as it now exists, possible, for necessity is one of the forces that makes the wheels of our society keep turning, to which I should add that we do not have any control over it, but ***IT*** does.

IT has taken me away from this game that does exist out there in order for me not to have to play this game that exists out there. There are certain things that I do have to accept, which are that what is happening out there. I should not try and change it, as if I was going to change things, or that it is up to me to make things happen, or to make things change. Also, I have to remember that everything out there does not belong to me, and that I should in no way think that I will take anything with me, for this would be a clear indication that I got lost again.

One more thing, in being with ***IT*** I am not winning or losing possessions, for I now realize that every thing belongs to ***IT*** anyhow. And that what I am gaining is the opportunity to continue existing as this living, existing moment that ***IT*** has permitted me to exist in.

We are not aware that ***IT*** is every thing to begin with, so it will be easy for us to get lost in this game that is taking place. ***IT*** is never around nagging you, yet ***IT*** energizes every thing that we do. ***IT*** never tells you if you are doing it right or wrong, because to ***IT***, it makes no difference what we do, for ***IT*** has a way of making happen what ***IT*** wants as a result or as an outcome.

Games

Regarding games, I have noticed that one of the most frequently used pieces of game equipment is the ball. Many games use balls, even as small as marbles, which I played when I was a kid and really enjoyed it. One thing about playing any game is that sometimes you win and sometimes you lose. The enjoyment is in the playing. After marbles, the balls get bigger; there's ping-pong, tennis, billiards. From there, the ball gets bigger still; we have baseball, football, and bigger still, basketball, volley ball at the beach, and then the ball gets heavier, you have bowling balls. And there are many more games with balls. One ball to stay away from, however, is the ball and chain.

Television

Here is another manipulation of the mind. I suppose when TV's first came out manufacturers and sales people did not think that the public would be walking around with measuring tapes to verify the sizes of the screens.

Let us take my case: I bought a 27-inch screen TV. The 27 inches are diagonal, so for me to enjoy 27 inches of screen I have to look at TV with my head tilted. If I did I would have to see a chiropractor every couple of days. The screen actually measures 22 inches across, and from top to bottom it measures 16 inches.

As for computer monitors, it just got worse. I thought I bought a 15-inch monitor, when I measured it I got 13¾ inches diagonally, 11-1/8 inches across and 8½ from top to bottom. From outer edge to outer edge I got 14 ¼ inches.

I did not give up, I went to see if the 17-inch monitor was better. What I found was that it ***measured*** 15¾ inches diagonally, 12-5/8 inches from left to right and 9½ inches from top to bottom.

But I did not give up here either. I went out to the store to see if maybe the sales person that sold me my TV monitor had taken me for a ride. And here is what I found with the more modern monitors, like the flat screens; the first thing is that they are still very expensive. As far as measurements go, the selling dimension is based on the diagonal, just like other screens. On a 15-inch laptop I found that it actually measured 15 inches diagonally, 12 inches from left to right and 9 inches from top to bottom. I feel that the most truthful thing would be to indicate how many total square inches the TV screen really has.

⌘~~~~~~~~~~~~~~⌘⌘~~~~~~~~~~~~~~⌘~~~~~~~~~~~~~⌘

My God, Please forgive me if I tire or bore you by thanking you so much.

❀~~~~~~~~~~~~~~~❀❀~~~~~~~~~~~~~~❀~~~~~~~~~~~~~~❀

Components

Here is another piece of manipulation. It has to do with your stereo system and watt ratings.

There was a time when you could buy a stereo system, it would be rated as delivering 50 watts per channel, that would mean that you would be having sound projected at a legitimate 50 watts through each individual channel.

Now when you buy a music system you will notice that it might say that it is a total 200-watt system and that you are going to get 200 watts divided by the number of speakers it has. This rating can be related to a right and left speaker box system. The problem is that each speaker box has three speakers, each individual speaker will deliver 33 watts. I asked a sales person if the rating meant that if I put all six speakers in one box, would I get 200 watts of power? His answer was that no, not really, the bottom line is that you get 33 watts per speaker, period. And here I was thinking I was purchasing a 200-watt power sound system.

Weapons

Here is a positive take on weapons.

We have to get ready to find the weapon for a big battle. This battle will involve the whole of planet Earth and should be taking place before the year 2880.

This is a negative becoming a positive. We have seen the negative effects of what came out of missiles and warheads when applied to human relations and politics. Yet it is this same negative that will be reshaping into something that will most likely save mankind on this planet.

It is expected that an asteroid, which has been classified as asteroid #1950DA, will come so close to Earth that the possibility of being hit is great. The U.S. government is already classifying this asteroid as cosmic enemy number one and is gearing up to find a way to destroy it. Among some of the options is that we launch missiles to destroy it or at least deviate it from what could be a collision course with us.

If this asteroid were to hit us it will have the power of a 100,000-megaton bomb. To make matters worse, there actually is talk that there may be more than one asteroid on the way. This is a situation that will have to be solved by the generations to come.

The crying syndrome

Imagine that there is a room full of healthy sleeping babies. One of them decides to cry to get attention, not because it is in pain or because it is hungry or wet, but rather, it cries because it wants attention.

This crying wakes up another baby who also starts crying for no reason other than that one is already crying. Before long, all the babies are awake and crying.

As far as they are concerned, it is normal and perfectly right to cry; and, this crying will result in getting something (attention) for nothing. This is what I call the crying syndrome. What we must not forget is that the most precious gift we have is the gift of life itself.

If we were not alive, we would not be able to cry when we were in pain. Therefore, in crying, we likely miss what is truly important, to be grateful that we are here in this moment as life, for it is a one-time gift that will never be given to us again.

Exercising the mind

Here is something to think about. From birth to death we are in the process of searching and reaching for a maximum development.

Once we are fully developed physically we have to continue to transfer energy to our bodies. One of the ways we do this is by exercising, which does not become any easier as we age.

This also applies to the brain. Luckily, it is easier to exercise this organ; all we have to do is think. A thought does not have any physical weight. The only part that is physical in the using of our brain is in the carrying it around in our skulls.

We know that our brain is physically smaller than that of our primitive ancestors[6]. Our modern day brain is smaller, with more tightly compacted capabilities; it is a brain that does much more thinking (exercise) than any of our predecessors. We have so much more information and such a more complicated environment today that our brains are much more agile and exercised than ever before in our history.

We must always exercise our brain, the older we get the more necessary it is so that we can avoid as much as possible the onslaught of diseases such as Alzheimer's. It is clear that challenging mental activity keeps the brain sharper and healthier.

The process of writing this book has been a growing experience for me. I think it can be the same for you, dear reader.

Right handed people

If right handed people are using the left side of their brains, then left handed people are the ones using the right side of their brain. It follows that left handed people are the only ones that are in there right mind. That's just joke…

⌘~~~~~~~~~⌘~~~~~~~~~~~~~~⌘~~~~~~~~⌘

**** ***IT is the ancestor of all ancestors*** ****

Opinions

When provided with an opinion, the person receiving the information is the one that has the ability to change. The person providing the opinion is merely stating a thought. When someone asks my opinion, I answer that I have none.

---

[6] The brain of *homo sapiens neandertalis* averaged 1450 cubic centimeters, while the brain of modern humans, *homo sapiens* averages 1350 cubic centimeters.

I believe that opinions are a way for the mind to entertain itself. For example: let's say that a 20-car collision occurred because of a drunk driver, and that it resulted in the death of ten adults and two infants. Opinions could range from claims that the driver should be put to death to claims that our roads should be safer.

⌘~~~~~~~~~~~~~~~⌘⌘~~~~~~~~~~~~~~~⌘~~~~~~~~~~~~~~⌘

*** ***It is easy for us to discuss what people should or should not do. Talk is cheap. People are going to do what they choose no matter what you and I think. It is much easier to talk about someone else than to effect positive change in ourselves.*** ***

Important here is to see that those opinions stem from an event that has already taken place; nothing done or said could change that now. The actions that took place resulted in some type of reshaping.

Instead of giving your opinions, accept and be grateful to ***IT***. Keep your eyes and ears open so that you may better understand ***IT*** as the moment, and are more available to accept your next role. Opinions require energy better utilized elsewhere.

The invisible governor

When we are voting for the governor and a bunch of other offices, I observe people making hay about casting a vote for who they feel should be the next governor. We humans have gotten used to the thought that we should have an elected official to give us leadership. This is natural; we want someone to give us the sense of heading into a better, happier future. We have always wanted this, I suppose.

I mention this because I have stopped voting. I am politically free. I do not need to vote for someone to lead me to a better, happier future. I have found a constant form of happiness. I used to live on a roller coaster where going up was exhilarating, and coming down was a drop into the low lows.

Throughout my life I have never observed a governor that could offer me a ride that was pleasant, peaceful, and most of all, enjoyable. You ask, what else is there? Well, here is what. I have a governor that is apparently invisible, yet you see him in everything, and you are in it. ***IT*** is the only invisible governor that ever existed.

Let us look at this invisible governor. ***IT*** governs every atom, everything you see.

Whoever we think we are electing to govern us is still ***IT***. The problem is that the people we elect forget who they are, or do not know that we are made of the pure energy that continuously reshapes the universe. If you don't know that and keep it in mind you actually start thinking that you can create something better. Well, even if you can, ***it*** is still ***IT*** reshaping.

I asked ***IT*** to take me away from the roller coaster I was on, and ***IT*** did. I am on a smooth, constant, peaceful road now. Some might say "well that sounds boring", but keep in mind that I can go back to the roller coaster any time I want to. But I am not.

I tell my friends that I do not vote anymore and they say that the system of government and everything else is necessary. I reply that yes, it is, but not for me because I know that all that is happening is ***IT*** in ***ITS*** constant search to reshape into all existing possibilities. Even all the tension and craziness is just exchange of pure energy, ***IT*** at play, making us change to what ***IT*** wants in that instant.

It is easier if we remember that we are talking omnipresent here, everything everywhere is ***IT***. And since we are programmed with needs and desires, we act accordingly.

I know it is not the politicians that are wrong, they may not know that they cannot make you happy and fulfilled, but I suppose they should try. But only ***IT*** can provide the happiness and fulfillment through the way ***IT*** governs our existence. Our Governor!

I will not change being close to ***IT*** in exchange for thinking that by casting a vote for some political figure I will be happier. I know that every atom of my body is governed by ***IT***, that will outlast all of us in this omnipresence.

Our desire to vote is driven by the idea that with ***it*** we can affect change. But I say that only ***IT*** serves me, and does so better than any elected official ever has or could. And I have never had to elect ***IT***.

So I hope you understand why I do not need to vote for an impersonal governor, when I have the most personal, greatest, and longest lasting governor that has ever existed.

Our environment is like a roller coaster. When our environment is on the upswing, we think that things are better. There is the old saying that what goes up must come down; this is how all things work. When our environment is down, we think that things are bad.

There is a better way of looking at this. The energy that exists within us is not an up or a down; it is a constant. This constant energy is level and peaceful. If we surrender to ***IT***, life can be much more balanced. To surrender to ***IT***, you have to except that you can no longer play life's games. There is no longer a winner and a loser; there is only an observer and a participant.

Freedom

I believe that freedom comes in different packages. There is no total freedom, especially for humans.

Since we have already established that we are created and that there are two primary programs that are going to take place during our stay on Earth, we are not totally free.

The first program that we previously discussed is survival. Humans need to survive long enough to complete the second program, which is reproduction. These two programs are necessary for ***IT*** to continue ***ITS*** reshaping process.

Males are required to impregnate. Women are required to give birth. As mentioned before, generally speaking, even the most liberated women have a desire to experience childbirth. Women have this desire so strongly that they are willing to do this in the absence of a male companion.

Most people feel that they are free to make the decisions in the area mentioned above. Human beings need to remember that we cannot create anything alone.

We are what looks like a creation that ***IT*** has reshaped into. The same applies to children. ***IT*** has allowed for human existence, ***IT*** has allowed for human feeling. ***IT*** has given humans the ability to love, hate, feel pain, and the ability to see and hear. ***IT*** gave us the gift of understanding.

From the beginning, I have been told that there are certain things that I have to do. It was not until I asked my GOD for help that I understood freedom. I did not expect ***IT*** to help me. Before I accepted this, I didn't consider the importance of God. Now I accept that ***IT*** came to help me. I asked for this help and have stayed close to ***IT*** ever since. Staying close has enabled me to see the gifts that are available to me. These gifts are never ending. Accepting ***IT*** has proven to be the best thing in my life.

The first change in my life after accepting ***IT*** was that I no longer needed an employer. I no longer needed to be told what to do, or how and when to do it. This is the first form of freedom that ***IT*** has permitted me to enjoy for the past 20 years as of the year 2005.

When a person is told what to do, they are denied their freedom. Everyday, this happens to someone. You do it; most of us do it. I choose not to deny others' freedom by hiring people that can do the things that I cannot, and determining the monetary energy required to complete the task.

We should be aware of alternatives and options. They give us the opportunity to improve the quality of our lives. There is always more than one answer to any question.

❁~~~~~~~~~~~~~~~~~~~~~~~~❁❁~~~~~~~~~~~~~~~~~~~~~~~❁

People tell me that I do not smile a lot. I do not think that you have to be smiling to be happy. As long as I am not sad and all is OK in my life; I'm happy. I call my state of mind being in a state of continuous contentment.

⌘~~~~~~~~~~~~~~⌘~~~~~~~~~~~~~~~~~~⌘~~~~~~~~~~~~~~~⌘

There is more than one way of doing things. I used to work alone by choice. Working alone I had to think of alternative ways of doing things. Working alone taught me to use the resources that I had available to solve the problems I encountered

IT has taught me that I do not have to work physically hard to survive. However, I accept the possibility that pick and shovel work may only be a moment away. I see my life as dangling on a very thin string that could snap at any moment. My life may turn upside down and for the worse at any moment. ***IT*** has taught me that whatever ***ITS*** plan for me is, that is the plan I will live.

IT is now my employer. ***IT*** has actually been my employer all along. ***IT*** is the best employer that anyone could ever have. ***IT*** has a way of letting you know what to do and when it is necessary to do it.

ITS plan cannot be seen by us, at least not until after the elements of the plan have taken place. I believe our inability to see ***ITS*** plan is a safety precaution that prevents human intelligence from interfering with ***IT***. We must simply accept ***IT*** and that in ***IT***, there is a plan.

IT does not force you into acceptance. ***IT*** allows you understanding, so you can come to ***IT*** on your own. **It** is up to you to accept. ***IT*** does not debate nor care about your opinions or philosophy.

Let me explain this with a personal experience. Before I could see what was happening to me, I fell into this hole called addiction. By the time that I realized what had happened I was in too deep to find my way out. I looked to the outside for help but found that people could not offer me a way out of my situation.

I asked God, from the bottom of my heart, to help me. I confessed that I was lost, and I did not know my way out. It was at this moment that I became aware that ***IT*** does listen.

IT can see, hear, and feel because ***IT*** created the human body.

A few days after asking ***IT*** for help, a big change started to take place. This change has lasted to the present moment. The change was noticeable right at the beginning, but I did not understand it.

After a few years had passed, I looked back at the many changes that took place and began to understand ***IT*** as I understand ***IT*** at this moment.

Life got better and better for me. My friends now ask me how I am doing. The first thing that I tell them is that I am alive, like you, and then I say, "I eat well, I sleep well, and I feel well; these are the best of times for me; and my cup is overflowing."

Every moment that I get a chance, I thank ***IT*** for allowing me to live. I see now that since I asked for change, I have received change.

❀~~~~~~❀

****** I have been given freedom as this existing moment. ******

⌘~~~~~~~~~~~~~~⌘~~~~~~~~~~~~~~~~⌘~~~~~~~~~~~~~~~⌘

Thank you

Thank you, dear reader, for having read this far. And while I am on the subject of giving thanks, I would like to say that for me, there are several kinds of thank you, each appropriate to a specific situation.

1- There is the thank you that I say to the supermarket attendant, when they tell me where an item is. This kind of thank you does not produce too much feeling inside of me.

2- Then there is the thank you that I give to someone for helping me at a particular moment. You might be familiar with this type of thank you. There are times when you need help from a certain person, and you can see are very busy with something that they too are busy doing, but they are willing to take a moment of their time to assist you, for they know that they have the answer to your problem. This type of thank you does produce a minor feeling inside of me when I turn to the person and say to them, thank you very much for helping me out with your assistance. You might have had this feeling for example when your car brakes down on a road that you are not familiar with, and you know that the last inhabited place that you passed on this road is very far away, and someone passing by stops and takes a moment of there existence to help out in what ever way they can.

I once personally helped a driver that had run out of gas on a road that very few people used, and I could see that that the thank you he gave me was a very warm one. When he tried to pay me for the gas that I gave him, I told him that it would be better that he keep his money, and that if at some point in his life he found someone that needed his help, to help them instead.

I remember another experience related to giving with out expecting anything in return. One day I was riding on one of these buses for which you have to have the exact fare. This man got on the bus only to find that he did not have the exact amount required. So he asked us passengers whether there was someone who could give him change for a dollar bill. I saw some people look to see if they could change the man's dollar, and so did I, but it seemed no on had the required amount, me included. I then asked the man how much he needed to complete his fare, and he said 25cents. So I gave him the 25 cents so that he could get on the bus. Then came the problem, for he said that he would ask each new passenger who got on the bus for change for a dollar so he could pay me back my 25 cents. But I told him that it would be better if he just remembered that if someday in his life he ran into a situation where some one else needed his services, that he help them instead.

3. Then there is the thank you, you give to someone who you have not even met in person, yet has all the qualities mentioned above, in addition to the fact that even though you have never met, they are willing to help you when you ask them, without any interest other than their kindness.

4. Then there is that very special type of thank you that arises when one personally asks ***IT*** for help, and ***IT*** answers, and you become aware that ***IT*** is there watching over your existence. Now, this is a tremendous feeling, because you becomes aware that you exist and ***IT*** exists, and that ***IT*** will always be there waiting for your call for help. So let me take this moment to say again, thank you (to ***IT)*** for letting me know that ***IT*** will always be there, just a whisper away. The feeling that I get it is like this: Imagine that you are in the middle of an ocean and there is no one near to help you out and you know that somewhere out there are sharks, and you know that this maybe the last moment that you may get before you personally go to see ***IT*** as death. Yet in your call for help, ***IT*** sends you a helicopter with the best ***IT*** has reshaped into, as the helicopter and the people in it exist as, who do not know you personally from before, or will not know you after this event, yet these experts will make sure that you continue to exist for more existing moments, as a gift from ***IT*** to you only. You know this so well that when you say thank you to the helicopter, and all its personnel, this thank you has a very powerful feeling, which can be described as the phrase goes, "from the bottom of your heart".

There's a lesson to learn here, as another proverb says: Always give and don't ask why, or to whom, and remember to never expect to receive from the one that helped you, but always be aware of the person that does need your help, for you should remember that we are all ***IT***. Some people might understand this last concept better if they remember this: God created every thing. While others might see it better this way: Every atom that exists, down to every atom we are made of is this pure energy (***IT***).

Who I am

Let me share with you a few more things that I have found related to this subject of who I am as the human that started out as that speck of life that was permitted to exist at a particular moment in Earth's rotation. Once I was here I would be the only one that would be traveling a particular road, only at a particular moment. Others have traveled on this same road, but at a different moment. Others will also travel it, and use the same seat in the same school, etc. What makes me unique is that I am the only one traveling the road as my moment. I am the only one adjusting and reshaping to my surrounding as this particular moment.

Since two things cannot be in the same place at the same moment, I am that moment that was permitted to come into life as a human. I say permitted because life is a gift that comes only as a result of this pure energy's existence. ***IT*** has allowed life to exist as a result of ***ITS*** reshaping, which permitted me to be here as a thought; because as humans, we too are omnipresent.

This would mean that you and I are 100% of that pure energy that is the Creator in which we live as an omnipresent thought. You can say that you are everything that exists and everything that exists is you.

But let us get back to that which we think we are. We can then say that you are all those roads that only you traveled on at exactly that particular moment in Earth's rotation.

Here is an interesting thought related to ***IT*** and life. You are here because you exist as life. Your life is housed within a body that started at conception, developed into a baby, and kept reshaping into maturity. But the life force that exists as you does not change throughout your life. It is constant, it does not reshape during your life. I find it strange in that life is the only thing that does not change. I have to take back having said that the only thing that does not change is change itself. I should have said that change is the duality to life that is constant as ***IT***.

Your inner self makes you different from others. You were the only person born in that particular moment in that particular place on this planet. You were the only one permitted to travel on those particular roads, attend those particular schools, and meet those particular friends. Sure, there were others on those roads, in those classrooms, befriending the same friends, but not in the same moment or way that you did. Your decisions and perceptions came from your inner self.

⌘~~~~~~~~~~~~~~⌘⌘~~~~~~~~~~~~~~⌘~~~~~~~~~~~~~⌘

*** ***Who are we? To the mind we are Tom, Dick, and Harry. To the spirit we are IT*** ***

⌘~~~~~~~~~~~~~~⌘⌘~~~~~~~~~~~~~~⌘~~~~~~~~~~~~~⌘

Instead, we attach ourselves to our work, our household, our material belongings, and even our children. We actually believe that our children be long to us. Even as adults, we believe we belong to our parents! A day will come, possibly not until we are dying, when we realize that we cannot take any of our possessions with us. We will realize nothing and no one ever belonged to us.

More about exercising the mind

While it is medically possible for a human to exist without the use of a mind, in this moment that I have come to understand as the here and now, ***IT***, has reshaped our minds into the most intelligent of all known beings. ***IT*** has allowed us to participate in what we call life. Some humans are even considered geniuses because of their minds. A genius usually has a mind that has not been damaged by say a blow to the head. So duck any flying frying pans to hold onto your intelligence and know when to duck! Head trauma may lead to craziness, which is the other side of genius. Being crazy does have a merit, however, as crazy people might feel more freedom to try what geniuses or normal minded people won't.

For our minds to achieve their maximum, they require an ongoing transference of energy. You would think that this organ would be easier to exercise than the others since thoughts do not weigh more than carrying the brain itself.

**** If you say that I am crazy, I am to you only. ****

Our brains are even smaller in this moment of our existence than they were during the lifetime of our primitive ancestors, perhaps because we have to think about more in order to survive than we did at the onset of our human existence.

We have exercised our minds with more information. And we should continue to exercise our brains by thinking outside the boundaries of our limited perceptions—perhaps this could help prevent Alzheimer's.

Fifty years ago, if you noticed saw someone walking and talking to him or herself you might think he or she was a little crazy. We see this today all the time; the difference is people are talking into cell phones. And for those who are a little crazy and want to talk to themselves without looking crazy, fake cell phones are available. Or, you could hold a microphone while talking to yourself and be mistaken for someone famous. Additionally, if you look like you are thinking without moving your mouth, you will look like a genius.

Meditation

Meditation is a way to let go of the attachments and become centered in the peace that exists within us as God.

When meditating, I am the eye of a hurricane. I need not control the spin of the storm around me. I float in the center where I am safe and peaceful.

It seems that most people do not realize the peace within until they come close to death.

****Only IT has been with me since the moment I became alive. Only IT will be with me when I take my last breath of life. IT is my best friend.***

Have you ever noticed that when someone is about to die, their eyes look up ward as though toward the third eye? People who meditate consider the area above the brows the third eye. This is the eye that sees within rather than without. This is the eye that sees ***IT***.

It was after a near fatal accident that I looked within to find pure energy as ***IT***. I asked ***IT*** to show me what I am supposed to do here on Earth. This is when I began to understand. This is when ***IT*** provided me guidance and clearer ways of understanding ***IT***.

IT did not provide me answers in black and white. ***IT*** does not send written messages. ***IT*** will not provide me life instructions through my mind.

⌘~~~~~~~~~~~~~~⌘~~~~~~~~~~~~~~⌘~~~~~~~~~~~~~⌘

****** Inner meditation is that place in which we can be at one holistically with IT, not as a religion, race, or color.******

❀~~~~~~~~~~~~~~~❀❀~~~~~~~~~~~~~~❀~~~~~~~~~~~~~~❀

IT communicates with me through an inner voice that I cannot explain. To my earliest question, ***IT*** replied: **"Do that which only you know how to do, and do it well, and do it only at the moment that you are supposed to do it, and do not ask why**."

⌘~~~~~~~~~~~~~~⌘~~~~~~~~~~~~~~⌘~~~~~~~~~~~~~⌘

****** O God, I wish that there was a way, that I could stay as close to you as possible, other then in meditation, or when I have to cry out to you for help ******

❀~~~~~~~~~~~~~~~❀❀~~~~~~~~~~~~~~❀~~~~~~~~~~~~~~❀

Maharaji

Prem Rawat, also known as Maharaji, an international teacher of inner peace and contentment, consistently reminds his audiences that which is most important exists within. He passed four techniques onto me that for the rest of my existence will enable me to connect with the pure energy that permits me to be here as a moment of life. They are simple techniques; they leave no excuse or need to travel to the Himalayas or stay with monks for years in order to know the energy that exists within.

Maharaji only asks that all who receive these techniques practice them one or two hours a day. And for those, like myself, who have physical limitations and cannot endure one-hour sessions, we can still do the best we can, and that is what I have done. I give fifteen minutes to ***IT*** twice a day, and they have been the best moments of giving in my life. They've led to my discoveries related to ***ITS*** existence as the pure energy within me as well as ***ITS*** existence as the pure energy that ***IT*** exists as the Universe.

*** ***You are not who you think you are. You are that infinite energy; you are IT, as IT evolves.*** ***

If we all gave just 5% of our lives to glorify this pure energy as ***IT*** instead of glorifying the human mind as to what it has done, is doing, and will do - we could always be in heaven.

Visit the Web site for Maharaji to learn more about his teachings**(<http://maharaji.org**>), but in my experience, giving 5% is largely a matter of remembering that ***IT*** is everything that exists, which is to say that ***IT*** is huge; yet even sleeping with ***IT***, we do not realize ***IT*** or understand that ***IT*** has always been with us.

******A thank you to M, for showing me that once you get close to IT there is nothing more important than IT.***

Religion

I have not mentioned religion yet. As far as this goes, I have none. Friends have on occasion told me that if I do not have a religion, then I must be an atheist. But I am not an atheist, for according to the definition, an atheist is 1) one without a God, and 2) one who believes there is no God. And as I have told them, while I do not need a religion, I do have within me that which we call God.

I do not need an intermediary between my GOD and myself. ***IT*** already exists within me. You can understand this better if you consider that everything is omnipresent.

We are God as ***IT*** reshaped ***ITSELF*** into the human form. Since we are God, there is no reason to have an intermediary. There is no reason for me to go anywhere outside my self to find ***IT*** since ***IT*** is omnipresence.

⌘~~~~~~~~~~~~~~⌘⌘~~~~~~~~~~~~~~⌘~~~~~~~~~~~~~~⌘

*** ***The kingdom of heaven resides within you.*** ***

~~~~~~

Me and IT alone

Let me share with you the reader some of the things that exist in my experience that relate to ***IT*** and me, and the best way for me to explain the situation between ***IT*** and me is this:

First, let me start with letting you know that I was raised knowing that there is a God, and that a church existed, as a place where I could go to be closer to ***IT***. However, as I matured I stopped going to my church, but I still believed in a God , that existed somewhere out there, and it was later on that I came in contact with Maharaji through a friend that took me to hear him speak at a Fordham University in New York.

After Maharaji finished his speech, I thought about what he had said and the only thing that I thinking to myself was that, what I heard was either too perfect or too crazy.

So I kept going back to hear what he was talking about, and let me mention that at that time Maharaji was not even a teenager yet.

What he was talking about was that there existed inside of each and every one of us, the most important thing that can exist, and that we as humans could connect with this energy that existed inside of us. So I asked him if he could show me how to connect with what ever is inside of me, which he did by teaching me 4 techniques that I could use to connect with my inner self. Le me mention that he shows these techniques for free, simply because what is inside of me always belonged to me, for he was not giving me anything; he was just showing me how to connect with that which was always there. So I tried connecting with my inner self.

However, I have to say that I am sorry that I cannot share my experience with you, because this is when I understood that what takes place between ***IT*** and me, will always have to be between ***IT***
~~~~~~

and me, for things get very personal when it relates to me and ***IT***.

I always tend to ask questions about things or situations, such as: "Why is this so ?"

So I started looking around to see what kinds of situations exist between people and ***IT***, and I found that what existed for me in my life as a God, also existed for most people, and that there were some people, like priests and monks that where trying to stay as close to God as they could possibly be, be it in a holy church residence or in temples, or ashrams, just to mention a few.

I also saw that there are people that could be considered perfect masters that were within themselves in the closest relationship with ***IT,*** that is, people that had no intermediates between ***IT*** and themselves. For it is in what we call perfect masters, that have no others in between, besides that person himself / herself and ***IT.***

Now, I am no priest or master, yet I find that there does exist a personal relationship between ***IT*** and me, which does mean that as individuals there can be a personal relationship between us and ***IT.*** In so far as the masters are concerned, I can see that these masters are here to show us individually how to be with ***IT***.

I feel great gratitude in that ***IT*** permits me to be able to be as close to ***IT*** alone. What I am trying to say is that I find that I am continuously thinking of ***IT***, about how ***IT*** exists outside of me; but as for what goes on between ***IT*** and me inside, that part will have to stay personal, for I do not see what happens between ***IT*** and me as something that can help other people, for as I have said, that relationship that exists between ***IT*** and me can only exist in that way, because it is something private, only between ***IT*** and me.

However, when I look outside of me, then I see that what is happening out there are events that ***IT*** is causing, and the reason why I say that these events are ***IT***, is because since ***IT*** is everything that exists, then everything that is happening is because ***IT*** is the mover and shaker, as omnipresence.

So my conclusion is this: The reason why events are happening out there is because this is the way ***IT***, as a positive and negative force, makes the changes that are necessary for things to go in a certain direction, so that ***IT*** can continue ***ITS*** reshaping.

I have accepted that I will in no way try to change what ***IT*** has planned.

This is why I find that as for my personal instructions in ***ITS*** existence are, as I have said else where, that I will do only that which I am supposed to do, and only as that moment of my existence,

For I now know that ***IT*** has always been ***IT***, as both the director and the actors in this play where humans exist, be it on Earth, or as we leave this planet, and ***IT*** will still be both the director and the actors, for ***IT*** is everything that is known as being created by ***ITSELF*** within ***ITSELF*** as omnipresence.

And as for you the reader, I am sorry to say that I cannot help you to be happier in your existence, for I have found that for me, the only one that exists, as being the most important is ***IT***, as how ***IT*** exists within me, and all I can do is to continue being an observer of what ***IT*** is doing outside of me, as ***ITSELF***, for everything as a totality is really ***ITSELF***, down to the last atom that exists inside of ***ITSELF*** as this universe.

Even a priest as a holy man, does not stand by himself, for he needs the companion of other priests or a pope. This is something that is not necessary for perfect masters like Christ, Buddha, or Maharaji, just to mention a few masters.

Spiritual growth and resistance

We are made in ***ITS*** own image, which is to take part in ***ITS*** ever changing self.

I say this because as we go into what we call bad times, such as a hurricane or an earthquake, that can leave us stripped down to only our existence, where we have lost everything as our possessions; it is during these disastrous moments, when ***IT*** takes just about everything we have away from us, that some of us are forced to come back to being closer to ***IT.*** Stripped of our possessions, which we should remember really belong to ***IT*** anyway, because it is a fact that we cannot take any of our possessions with us when we die, I have observed that in these moments of disaster, some people will say, even after they have lost every thing, "Well thank

God, for we are still alive", but others will continue to cry and protest that they have lost every thing that they had. These who cry and protest forget that everything they had belonged to ***IT*** anyway and that the only one that can really help them is just waiting there for them to reestablish a deeper connection, between ***IT*** and them, as in when we say to ***IT*** : "**O God please help me!** I say it a little differently**: "O God please help me if you so desire."**

We should always remember when a disaster occurs that whatever help we may receive, is still ***IT***, for ***IT*** is the mover and shaker. You might see it better this way: It was ***IT*** that produced the earthquake or the hurricane, so that no matter what, how much we may want to complain or protest, as to why ***IT*** became the earthquake, or the hurricane, it won't help undo what ***IT*** has already reshaped into, as cause and effect.

I have personally found that when I have needed help, I first turn to ***IT***, for the help that will come has to be sent by ***IT***, for ***IT*** is every thing that will be sent to me by the humans that ***IT*** exist as, be it food or housing or clothing, etc. that will come to me as help.

Gifts

In life, there are different types of gifts that we receive. There is the gift that you will never, or hardly ever, use such as the Halloween candle or super sized fruit dehydrator, which you have to find a place to store for as long as you can. There are gifts that last only a short duration: flowers, candies, stationary, cologne, and there are the gifts that you do not like, or need, so you pass them onto others as a gift from you to them. These gifts could travel around the world and back depending on the connections of friends.

There too is the gift of fruitcake that you know you will not eat because if you do, it will put two pounds on you, so you also pass this extra weight onto someone else. Another gift we give and get is the one we are to share, the vacation cruise, the back massager, the backgammon game, and the computer software.

⌘~~~~~~~~~~~~~~~~~~~~~~⌘~~~~~~~~~~~~~~~~~~~~~⌘

*** ***The more you give during your stay on Earth, the more you will receive. Giving is the transferring of energy.*** ***

❀~~~~~~~~~~~~~~~~~~~~~~❀~~~~~~~~~~~~~~~~~~~~~~~❀

And on the subject of gifts, from my personal experience, I have noticed that sometimes friends give me gifts without knowing what I really need. In order to not do the same thing, I decided to pay attention when a friend tells me that he or she likes something. Then I get them precisely that item. However, this can also be a problem because I once ended up with not one, but 3 binoculars! I feel one should think up a way of finding out what a friend really needs without him or her becoming aware of it.

And then there is the gift that is meant only for you and intended to last until your last moment as life. My gift came freely to me by AA in the form of a group of people who had gone through what I had and even worse. They gifted me by showing me how I was going to control my addiction to alcohol and return to a healthier way of existing.

As a member of AA, I can confirm that the program is very effective in the help provided as people share with other people who want to stop drinking, and as a free service, I say this because I was a heavy drinker. I have noticed that most Alcoholics have a big demand for the consumption of liquids be it for coffee, soda, and the many other beverages. I have not had a drink of alcohol for what has permitted me to see more than 7000 appearances and disappearances of our solar sun as a healthy person that does not need a drink. I measure the time as over 7000 sunrises and sunsets because I do not keep track of my life as days, months, or years.

*** ***Be grateful for your existence. Be grateful for ITS existence. Trust in IT, for IT is more than a master at what IT is doing.*** ***

Think about this: Ever since birth, we have grown bigger, perhaps fatter, and even wiser in our reshaping, but we never left this existing moment. Since birth, we have only existed in one place: the omnipresent, the "here and now." And because this moment is the omnipresent that has always existed, even before the Big Bang, no matter where we have relocated our housing or even if we were someplace else in the universe, even if we had flown to the Moon, we would still have had to exist in this place called omnipresent, the here and now.

I know that every thing that I do, I can only do as this living existing moment, for I know that I was born in this existing moment, and I will die in this same existing moment, so I know that every thing I do has to be done within this same existing moment that I call the here and now. Therefore, as an AA member, it is not that I will not have a drink today; it is that I will not have a drink as this moment of my existence, and I will enjoy this moment as much as ***IT*** will permit me to.

IT is the only Master

The reason why ***IT*** is a Master, is because ***IT*** knows what ***IT*** needs to reshape into. ***IT*** is a Master at reshaping ***ITSELF***. Yet there is no perfect plan; we have no idea what ***IT*** is going to reshape into next. Even as you read this, ***IT*** has already reshaped. I trust in ***ITS*** reshaping, yet, ***IT*** remains mind boggling. If ***IT*** is everything including us, where does ***IT*** exist as the place in which ***IT*** exists?

And as much as I have looked around at the things that we do as in the way we as humans run things. I am very glad that ***IT*** is the one that runs this whole universe, from that atom that may find itself at the end of whatever this universe is. To that last atom that makes my existence possible, for even as I look at all the things that we think we are doing. I must never forget that long before I got here and my parents and grand parents, all the way to when we began to exist as humans, that ***IT*** was ***IT*** from the beginning, and that whatever ***IT*** wants to do with us as ***ITSELF***, as ***IT*** prepares to take us off this planet, to some other part of ***ITSELF***, we really have no say. For ***IT*** is the only one that knows what is out there as ***ITSELF***, and as for me. As this moment, all I have to do is enjoy as much as I can as ***IT*** keeps reshaping as this moment called life. Even if what I see outside of myself looks like madness, I have to continuously remember that ***IT*** knows why.

My job and my Boss

Let me give you a description of what my job consists of, and some things about my Boss that some of you might find interesting.

I start by saying that I would definitely not trade the job that I am now permitted to do. My job starts out as not having a real schedule, as a time system. The things that I must perform do not consist of a routine.

The meaning of work, when related to ***IT***, is the transference of energy so that something will happen. As a result I make something else happen, directly or indirectly. This way, ***IT*** can continue its travels as the nothingness of outer space.

Let me explain it this way: We are the first living things that ***IT*** has reshaped into as human bodies, that have taken its weight as atoms and reshaped them into all the things that we have produced since we have existed as humans, (due to the fact that our survival programming forces us to work out of necessity to produce the innumerable things that only exist on this planet, that come from ***ITS*** weight being reshaped as the atoms that we have been using to build everything that is not naturally made by ***IT***, as a reaction to the way ***IT*** does things, like the way it produces atoms using a force like the Big Bang, so that ***IT*** could then reshape into us), so that now ***IT*** can reshape itself in ways that perhaps ***IT*** has never done before. For not even rockets, computers, or canned food could be produced as a direct result of how ***IT*** did things before, such as the reaction that ***IT*** produces; for instance, like the Big Bang.

So you see that what we may call the things that we have done as work, from as far back when we started factories to produce things, is really ***IT*** down to the last atom, as the raw materials that we have used, and it is ***IT*** as the only boss that was here then and is here now, and will continue to be ***IT*** after we leave. And to this let me add that I have no problem accepting ***IT*** as my Boss.

Remember this: You have worked for other bosses without caring what the result from your work would be, and you probably couldn't care less about who your boss was, or is going to be.

I am very grateful to ***IT*** for the work I have been permitted to do, and I hope that ***IT*** keeps me as close to ***IT*** as ***IT*** has until now.

Now let me also share with you the meaning of having a boss, which is someone that will tell us what we are suppose to do.

Most of us understand what I am referring to because it is a natural thing, being told what to do, that most of us know, which is our acceptance of the presence of having a boss as part of our existence. We have had bosses that were so bad, to the point where we continue trying to find other bosses in other jobs that would be more pleasant.

I too have gone through this process, and by luck (for I was not aware of this luck that I have) I asked ***IT*** for help (I did not ask for change) with a situation in which my friends could not help me because they were of the problem. What I found was that in asking ***IT*** for help I was opening a door between ***IT*** and me that I had never used before. I saw that I had not been aware of how close ***IT*** really was to my existence, which you will read more about in different parts of this book.

~~~~~~

**** *Prolonged work can mold your mind and body* ****

And I should also add that it took many of Earth's rotations for me to understand ***IT*** better.

Now I can see that ***IT*** is my boss, to which I have to add that ***IT*** always was, from the moment that I came into existence. It is just that as we start out in our existence we are not aware of ***ITS*** direct connection to us. It exists right from the beginning of our existence that starts with our birth, as all the pure energy that we exist as.

So that I, like other humans that ***IT*** exists as, that are like me, were not aware of this situation. As I grew up all I saw was humans that thought they where the chief commanders as that moment.

That is why I say that I was lucky to have found ***IT*** as being totally everything that is out there; and as every atom that I am composed of.

At first I was not aware of ***IT*** as my Boss because I was still transferring energy as work in order to maintain my first programming, which is survival.

I continued to work, with the difference that I was working for myself (as I thought) as a concrete water-proofing consultant. This was a way of existing; it still is, but now as a service, not a necessity.
~~~~~~

But let me get back to my other job, for I still have a job where I have to do things (transfer ***ITS*** energy), but as the job I have with ***IT***, I do not really know what I am suppose to do.

Here are some of the things that I have found related to ***IT*** as my Boss. The only instructions that I have are these: do the things that present themselves as the existing moment, like taking care of the people that knock at my door for help, and answer the calls that people make to me. But never asking myself what is in it for me. I have learned that ***IT*** has given me everything that I need, that are my necessities, and a little more.

And as for the monetary energy that I get for my services, in transferring my energy in exchange for the problem solving services that I can offer, is not or ever was where I could feel that I no longer had to work, transfer my human energy.

I was born into a family that had to work for a minimum salary, a family where both parents had no reading or writing abilities and never made it as far as high school. They did not have an education.

From the beginning of my existence I knew that if I wanted to at least survive I had to transfer energy, as work, for whatever I could receive as money. I have accepted from my beginning that I was not rich in money. I could not do whatever I felt I wanted to do as work.

And as of this existing moment, to which I have become accustomed, I still receive monetary energy for what I do as work. With it, I at least cover my necessities. Like many others, I buy my groceries for a few weeks in advance, to ensure that I can at least make it that far.

But I know that ***IT*** is there, and that ***IT*** has never let me go hungry, or without a roof over my head so as not to get wet. I have gotten accustomed to the way our relation exists, for ***IT*** still feels there is no need for me to be a millionaire, to where I no longer have to work. I have become accustomed to the amount of monetary energy that ***IT*** sends me, to which I have learned from, because I know that, as an example, I receive a thousand dollars, I know as an expression, how full my gas tank is, and how far I could go on this tank of energy, and how far I should not go, as in doing something, or buying something before ***IT*** sends me more monetary energy.

Let me add that ***IT*** has never given me so much monetary energy where I could get totally lost. This relationship between ***IT*** and me has helped me understand the things that I can do, and what I cannot, or should not do; for as I have said else where, I know that up until now ***IT*** is worse than me for putting things on paper so that I could read what ***IT*** wants me to do. But since I do understand ***IT*** a little, of how ***IT*** is, and how ***IT*** operates, I know that sending me written messages would not work either, for ***IT*** does things only as an existing moment, so that by the time ***IT*** prepared me my written instructions to my does and don'ts, it I would be too late to use these instructions, for ***IT*** would have already changed into something else where the instructions could not be used or applied.

So that If I want to know what I am suppose to do as work in exchange for ***ITS*** energy (money) I have to keep a constant connection to what ***IT*** wants me to do, or not do, as an understanding that exist between ***IT*** and me, and not me and you.

This is why you the reader could now understand why when I started this book I never really knew if it would reach you as a reader. To me it made no difference if it was ever completed, and what the spin off from this book could be, because it was more interesting what I was learning about ***IT***, and me and ***IT*** and you.

This is why I say: "Oh Boss, please never dump me, for I have always done my best in serving you and all that you exist as that is outside of me as ***YOU.***"

But let me get back to my work, for my relationship with ***IT*** is as you can see, something that can only exist as being personal. Let me also add that when we do work, as work for others, we will also learn from the work that we are performing, and from the boss that we may be working for as that moment, be it as a factory worker or as an executive, and that sometimes in the work that we are performing as work, is not helped by the others that work there, for they feel that if they teach you too much, their job as work is in danger.

I have said the above, because my boss (***IT***) has always let me know that ***IT*** will not fire me for being absent, or for doing something not well enough. ***IT*** just says "do your best" in what you know how to do, and do it well and don't barter asking for a raise, for you already have the most precious gift that can exist, which is being alive. All I now have to do is keep up with ***ITS*** ever-changing self.

And as for the things that I do as work, they are not things that I have to do out of necessity. They are more like; well let us do this one as this moment, as doing something as work that will make something better than what it now exists as.

As for possessions, I buy that which I feel I may need to exist more comfortable in my everyday existence, but not to store as luxury or as something to exhibit. I have to remember that everything that I buy, or get as possession, I cannot get attached to, for they really do not belong to me, but that I can enjoy them only as an existing moment because I have to be aware that as possession, I will have to leave them when I leave as death.

Everything belongs to my boss (***IT***), so that I am told by ***IT*** "Do not worry about possessions as things, just enjoy them when you can and do not get too attached to them." ***IT*** will continue sending me the things I will need in order to eat well, sleep well, feel physically well, even though I will need to see what ***IT*** exists as doctors so that I will make the necessary adjustments to my body so that I can stay for more Earth rotations with ***IT*** and enjoy and understand ***IT*** better. This will also require that I continue thinking well.

So as you the reader can see, I hope ***IT*** will always be the only boss I have.

****** Thank you (IT) for permitting me as this moment of my existence as life, to answer to YOU only ******

Who is # 1

From our birth we begin to use the number one to designate what is most important. Many of us will say that our job is number one, that our tribe or family is number one, or that our home is number one. Many say that they are number one. Actually, ***IT*** is number one, figuratively and literally. Everything else is less than one. If you were to ask me where I would rate my opinion, my car, the president or a political party, I would have to say at zero. These things have no value. The more we consider these types of things priorities, the more we diminish and ourselves those around us. If we rate ***IT*** as number one, everything else falls into the appropriate priority as the power that zero has.

Impossible gift

What is the only gift you cannot give to the one you want to give it to? I have tried many ways to give something to ***IT***, but I always fail. It is impossible to make a gift to ***IT*** because ***IT*** is all that exists, including me and any gift I could possibly think of. How do you make a gift to something that is everything? I have tried in vain.

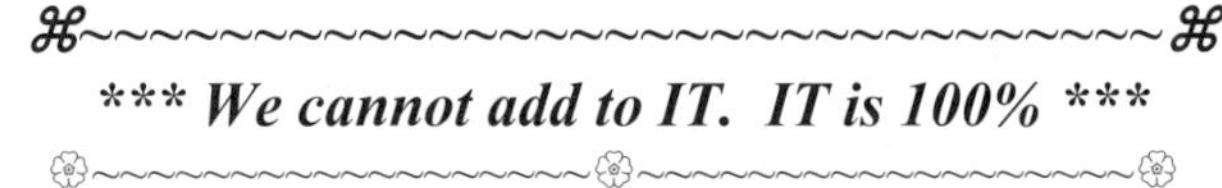

*** We cannot add to IT. IT is 100% ***

IT teaches us.

The next time that you are watching TV, all the programs and movies that are possible; remember that ***IT*** is showing us some of the possibilities available.

Resisting

Let me explain this with a personal experience. But first we should remember that we are built to continually grow, getting stronger mentally and physically. As a consequence we get into routines which tend to mold us.

I say this because it took a hurricane for me to become aware that as much as I have been very well protected, there are moments when ***IT*** will box with me. During a hurricane ***IT*** kept knocking me off balance, so much that I had to get on my knees and repair many things. In times of disasters we will find that there are moments when there is no one else around.

I became aware that ***IT*** will box with us so that we can rebuild those qualities that ***IT*** has incorporated into us as survival instincts. This will take place from time to time for some of us, and it is in these moments that we should never forget that everything that is happening is ***IT***, and that the only one that can help us is ***IT***.

I say this because it has become clear to me that in these disastrous moments I cannot blame anyone, for I know that we are also the result of violent events in ***ITS*** reshaping.

When I noticed that ***IT*** was boxing with me I became aware that the message was that it was necessary for ***IT*** to knock me down so that I could get up again. In doing so ***IT*** reinforced that I was neglecting the physical and mental abilities that ***IT*** has placed in me and all of us.

So that every time ***IT*** knocks me down, ***IT*** says get up, for you still have the most important gift of all, which is life! And also remember that ***IT*** is the champion and you must learn from it.

It may take a few falls to reinforce what you already have, but you should always remember that there are others who are in worse condition. If harsher lessons are needed, one can easily be put in worse situations.

Stress

Stress is when we try to do too many things within an existing moment. An example of stress is a jammed typewriter. If you typed the keys of an old typewriter too fast, they jammed together.

Speed requires space, if we go too fast, the space needed to accommodate the performance is not there. Remember the saying, "Haste makes waste," or in this case, stress.

The stolen crown

Here is one that I have tried to resist writing; it has to do with doctors. I hope that the doctors that I have to visit won't see this section. Naturally, this does not apply to all doctors in all countries, for I am sure that there are countries where it is worse, and in some, this situation may not exist at all.

If we look back into the history of doctors we will notice that there existed a moment when doctors where summoned to appear when the royal king requested them to be present, and when they where summoned, they, as doctors, had to respond as quickly as possible, for doctors then knew that when the king requested their service they had to come or they risked death by order of the king.

After the virtual elimination kings, doctors were more of a service, especially in war times, where doctors had to be an all-around provider as the service they performed. As doctors saw that there was more money in specialization, more fields of services opened up. As people aged and need specialized attention we began see more doctors for arthritis, heart doctors for high blood pressure, and so on.

Some people out there may have gone through what I will be mentioning. There is a group of doctors that won't give you an exact time for an appointment; it works better for them to say that appointments are in the order of arrival. To use an example, the doctor arrives at his or her office at 9:00 AM, but the office opens at 7:00 AM, which puts the personnel to work sooner so that the paper-work and the collecting of your money is done before the doctor arrives.

For you to see the doctor as early as possible you do have to get there before 7:00 AM, and if you are the first there it is up to you to make a list of the people that arrive so that when the doors open you can at least be number 1, 2, or 3, and sometimes #15, and if you get there at 8:00 AM you know you will be there for a few hours, or you will see the doctor after 12:00 PM.

We do not necessarily have to be sick in order to participate in the above described situation, because just for us to maintain our best, as being physically well, we have to visit these doctors just for our yearly check-up. And some doctors really look forward to you having a serious problem so that you will have to come back again and again; something like having a sure customer.

You will understand why I refer to doctors as those who have stolen the crown, for if you recall, the history of doctors was for them to service the king at His Majesty's convenience, and as time has gone on they now exist as a service where they are the king and you go to them when they say so, for they are the ones that are wearing the crown; they are the kings at this moment.

As I see it, they to will have to participate in the never-ending cycle of change. We can already see some changes, for there are now doctors that can offer some of their services by cameras through the internet to reach other people in other countries. And there are countries that are giving more attention to training more doctors to service their ever-growing population.

The line up

Those of us that attended school in the USA have been photographed at some moment for the yearly school album. This was likely the first mug shot ever taken of us. We did not have a number attached to us, but there was a number attached to the school year book, all in the name of education.

A birthday greeting

Congratulations on your birthday!

Let me start with why you should be grateful:

1-you are breathing, hopefully well, since this is priority number one for your existence

2-you have a water faucet nearby, can't live without water

3-you have a place to eat, at home or out where someone will serve you in exchange for monetary energy

4-you feel well because you rest well when this pure energy also known as God takes away this solar light which we call daylight and lets us feel its absence, which we call night.

5-you have the opportunity to reach your maximum potential.

Remember that since you have been permitted to be here in ***ITS*** own image, you have been given the chance to take part in ***ITS*** dual way of transferring energy. That means that you will take in and put out energy. That is the way that pure energy works, constantly transferring.

You have seen the duality of this transferring of energy in the many things that have been put in front of you as you have reshaped from a child into an adult.

You have been permitted to enjoy this gift called life.

We have learned since we were born that every 365.25 days Earth makes a full rotation around the sun and this makes a year.

Some of your friends will tell you that you are now 25 years old (we should be so lucky). But I would rather tell you that you are not 25 years old, I would prefer to remind you that you should be grateful that you have seen 9,131 Earth rotations, knowingly or not.

From where you have seen yourself from what you can remember as being a child, to what you exist as this moment has permitted you to do all the things you have done, from being that pretty child up until now. You might still be that pretty child.

During all those rotations you have had the gift to feel, touch, smell and see all the other things that go with being alive. You have been allowed to make love, be loved, see different streets, different people, colors and languages, clothing. You have known the many nice things that have been said about you.

You have tasted different foods and you have been permitted to experience brothers and sisters (hopefully).

And most of all you have been given the chance to connect with that God that put you together as a place that exists within you.

All of the above mentioned activities have taken place as a moment of ***ITS*** existence in this place known as omnipresent.

Diet, exercise, and health care

Dear reader, I am placing in this section subjects that have to do with taking care of this wonderful gift we have that is our body.

I am aware that many of the things that I will mention may not be of use to some of you. But I bet there are people out there that could make good use of some of the things I have learned while I have been dieting and learning to take care of my body.

Weight

From the beginning of time, the human body has been conditioned to gain weight; It is a normal and natural process. Weight gain is a safety mechanism. The storage of fat in the body guarantees the necessary fuel and energy it needs to maintain daily operations and survive until the body reaches maturity.

During our primitive development, excess fat storage gave human beings a better chance of survival, as the body will burn fat before it burns muscle. This can still be seen in developing countries where food is not as abundant as we are accustomed to.

Developed countries have an abundance of food and drink and plenty of waste, eliminating the need for the body to store as much fat. It is easy to see the effects of overabundance by looking at the obesity around us. Food transfers an incredible amount of energy to our bodies. We apparently have not learned this, as we generally eat much more than we need for survival. If we still had to worry about surviving without food it would not be a problem.

But as it is, 90% of us will gain weight over the years.

Weight accumulation is sneaky. If you consume an additional quarter of an ounce a day, you probably will not notice it on a day-to-day basis. But multiply that extra quarter ounce of food a day by ten years and you end up with a weight gain of fifty-five pounds!

To prevent obesity we must develop healthy eating habits early in life. One way to prevent weight gain is to reduce your daily food intake by a quarter ounce or more. It's not enough to affect your pleasure of eating, yet it would make a big difference in maintaining a healthy weight as the years pass. Of course, you can always let the weight accumulate then try an extreme diet. You might even lose the weight, but it's likely only temporarily, as your metabolism would slow significantly in an effort to survive.

According to current scientific dietary measures, one pound of extra weight is equivalent to 3,500 calories. This means that a 100-lb body has 350,000 calories as stored energy. For a 100-pound person to lose weight then, he or she would have to take in 500 less calories per day. This should result in a loss of one pound for every seven days. The body will not initially notice a change this small; therefore, it will not fight to survive. The body will actually get used to losing weight.

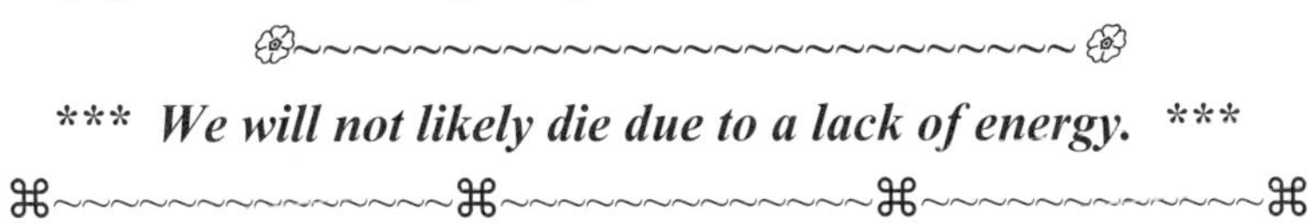

*** ***We will not likely die due to a lack of energy.*** ***

Resistance to losing weight

If by dieting you obtain a lesser weight, you may find it difficult to maintain that weight. The same 1000-calorie diet that enabled you to arrive at a weight of 100 pounds, for example, will eventually cause an accumulation of weight as the body's survival mechanisms create a safety reserve of energy.

This happens as your body's digestive system and organs, your intestines for example, accustomed to the fullness of more calories, function in the image of ***IT*** and strive to reach a maximum in development.

When the dieter then reaches the desired weight, he or she may have to increase the caloric intake so as not to continue losing weight. When I am at my ideal weight, I take the opportunity to feast at Sizzler Restaurant, which has a buffet of meat, fish, fruit, vegetables and desserts.

I would never eat there while dieting, but during my feast… well, it is the place to go when it comes to all-you-can-eat. And I can assure you that I am good at eating a lot, perhaps more then normal, which may actually be typical.

As for what we consume as food, my feelings are that our body uses 5% to 10% of what we eat and the other 95% becomes waste that ends up in the toilet. If there is a reader that has statistics on this and would like to share them with the rest of us, please e-mail it to **omnipresentrr@hotmail.com**

Now that we are aware of the body's reaction to overabundance, we should try to establish our eating habits as early in life as possible.

So, even if you have reached, for example, 100 pounds, your body will resist staying at that weight. Combine that with the problem that as we get older our metabolism slows down, the same intake amount of calories with which you obtained your ideal weight will try and accumulate as emergency reserve bodyweight.

I also noticed that when I was losing weight I would feel just a little hungry at the end of the day. This hunger came from my body having to use up its reserve in order to continue its normal daily activities, since I knew that this was going to take place for a short while I would just bear with it.

I also noticed that when I ate more at night I would not be as hungry in the morning. This was because I still have stored energy from the night before.

So it is better to eat a higher protein diet in the morning, and before you eat again you should try and hold off until you are hungry in order to use up some of your reserve.

I am now trying to continue to drop my weight to 125 pounds. At this point I can still see areas that have fat, I am trying to reach that weight at which I will have no fat attached.

Here are some things I found when I got to 128lbs. I stopped evacuating on a daily basis, which had been my past rhythm. As I got closer to 125 pounds I started skipping a day as evacuating, which tells me that my body was trying to keep that weight. So I then started weighing myself every other day, because the difference was 3 pounds.

Also, I had to eat less as I loss weight, for the quantity of food I ate to reach 130 pounds became too much to if I wanted to continue reducing weight.

I notice that when I eat at Sizzler it becomes more difficult to lose the next pound; it takes longer. I also notice that I get hungry sooner. For example, I used to eat my heaviest meal at 4:00PM, now I eat at 11:00AM, which leaves me a little hungry by bedtime. Another thing is that this bedtime hunger disappears by the morning.

This may sound facetious, but it would be great if there were a tapeworm that we could swallow that would eat the excess away until we reach our desired weight. Of course there would have to be a way to remove it (the tape worm) without it being one great big headache.

The more you have the feeling of being hungry the more it is an indication that you are losing weight. Remember though, real hunger only happens when you are actually losing weight.

When I reach the point where there is no fat I will inform everyone, this is a very slow process.

The good thing about dropping down to 120 pounds is that I can make at least 10 visits a year to Sizzler, where I can enjoy some nice big meals while climbing back to 135 pounds, which is my normal weight.

Keeping the weight off

Let me offer you my personal findings in relation to dieting. When I became overweight I started to analyze the situation, and I noticed that my problem started with my mouth. So the first thing I had to do was come to an agreement with my mouth. That was that what we have been doing was getting me into physical problems.

It was my mouth's turn to be disciplined. My mouth accepts being disciplined, I noticed it with alcohol. If I did not have one drop of alcohol I knew that I could lick the problem that follows after the first drink. And let me also share with you, a few more things that I found that are related to the mouth, or at least mine.

That is that it always wants just a little more, and that it is hard to satisfy the mouth when it comes to small quantities.

I can now understand why AA offers coffee as a beginning to becoming sober. Coffee helps you stay awake and alert, which is the opposite of being drunk. But what I really wanted to say is that I noticed that I was drinking a lot of liquids, which I still do, something like two gallons a day between my consumption of coffee, tea, and the liquids that my meals contain. It has helped me lose and keep off thirty pounds since ten years ago.

When I am preparing my meals for the day I do it when I am not hungry. I prepare fixed quantities for what I will be eating that day.

So we, me and my mouth, have come to an agreement, which is that it (my mouth) will only eat what I have prepared for each day. As a prize, my mouth can go out and feast from time to time. Actually, I need to because sometimes I drop below my ideal weight.

I have also notice that I cannot buy sweets, like fruits and granola, or cakes. When I dropped below my recommended weight I told myself that I could reintroduce some sweets to bring my weight up again, and this is when I noticed my mouth becoming undisciplined again. Like the old days with alcohol. AA has a saying that goes "One drink is too much and a hundred is not enough." For me, that also goes with sweets.

When I tried regaining a bit of weight, I bought cheesecake, granola, and fruit. Then I wanted to get them in bigger sizes. Then I had this big struggle at home just trying to deal with portion size. My mouth just wanted bigger and bigger pieces; it was ridiculous.

I noticed that if the stuff was not in my home, my mouth would not have that much of a problem. You know, out of sight out of mind. I stopped buying sweets and have maintained the discipline. I will add that for other people the big weight gainer food might be something else. For me it is sweets.

The upside for me is that I indulge in Sizzler when it is time to gain weight. I have a feast and revel in it for the moment. Then I go back to the discipline of the diet. My mind and mouth are at rest. It is all part of my maintenance program, the balance I have found.

The good thing is that I can enjoy the feasting out with friends, and it is all working for my betterment.

Your future weight

If you would like to know what your future weight might be, here is what you should do. Let us say that when you where 25 your weight was 100 pounds, and that now you are 35 and your weight is at 111 pounds, which doesn't mean much as far as your present appearance if you continue with your eating habits.

This is what happens to most of us as a natural way to gain weight. Let me share with you an observation. I have observed normal, everyday people in foreign countries, and they also have weight problems just like you and me.

I say this because we in the USA should not blame fast foods for our being fat. This is happening world wide, and like I have mentioned before, it is basically that most of us are designed to store energy as weight. The slowing of our metabolism after the age of 35 is also a contributor to this problem, for it seems that our body is programmed to slow down our metabolism, which contributes to storing energy as weight.

But let us get back to predicting your future weight. Try and recall what you weighed 10 years ago and then keep adding this difference to every 10 more years. If you weighed 100 pounds at 25 and you now weigh 111 at 35, you can more or less predict that at 45 you will weigh 122, at 55 you could weigh 133 pounds, and at 65 you might weigh 144 pounds. Now, at this stage you most likely will be having problems with your health, for you have gained 44 % of your weight since you were 25.

I feel that it is more important to dedicate some of my existence to continually monitoring my weight and continually disciplining my mouth. I now realize that this gift that I have as a human body has to be kept at its best by me, and not so much by others, like doctors. And like I have replied to friends that have commented that I now look skinny; I prefer to look slim and be healthy and well, then to be fatter with health problems.

More importantly, I can now visit Sizzler more often. I just do not like to eat out alone, so I invite single friends to eat out with me. Like I have told my friends: Getting fat is not the problem, all I have to do is eat more; especially sweets. And I know that in my living alone I can control my weight much easier than people who have to eat as a family.

I have offered this information so that you can look into the future and see if there is a problem coming.

This bears repeating: Weight accumulation is sneaky. If you consume an additional quarter of an ounce of food a day, you probably will not notice it on a day-to-day basis. But multiply that extra quarter ounce of food a day by ten years and you end up with a weight gain of fifty-five pounds!

❀~~~~~~~~~~~~~~~ ❀~~~~~~~~~~~~~~~ ❀

To better enjoy life, you need good health.* **

You should remember that this ¼ ounce is equivalent to 54 calories, (because it takes 3,500 calories to add one pound of body weight) for as strange as it may seem, it is these extra 54 calories a day that will get you to add 55 pounds in ten years.

But there is a chance you can start heading it off now by reducing just ¼ of an oz of food daily, in foods like fats, sugar, and carbohydrates.

You should start taking care of your gift. You should also remember that, like I found out, after your partner (mate) leaves you (if this were to happen to you), you will be faced with this over-weight problem which you will always have to deal with on your own, for it is your problem and it is up to you to control.

Let me mention what inspired me to write this piece. It was because a friend mentioned that I was looking skinny, and she was on the fat side. Since I knew her when she was skinny, I knew that she would continue to get fatter. But since I am not here to tell anyone what they should do with their lives, all I can do is offer what I see, for I have noticed that people can see other people's problems and comment on that rather then seeing their own problems and find ways to deal with them. It is easier to say what others should do rather than to do the things we should do to ourselves. This is not very logical, for I know that it is easier for me to make my own corrections because they are within my reach, and more possible than changing others.

I guess it is easier to say that people should change than to come to an agreement with oneself. Anyway, it is up to us to attend to our bodies as much as possible.

One pound a day

Here is something to think about. Let us say you weigh 100 pounds. Now, it is known that to maintain a body weight of 100 pounds you are required to consume 1,000 calories daily.

In my case, I dropped to 120 pounds to see where any fat left behind would remain. Like most people, that turns out to be in the belly area. I actually wanted to keep losing weight until I lost all residual fat but my friends told me I was getting way too skinny. So I decided to stop losing weight and started looking at calorie counts on food products, especially peanut butter.

For instance, the last time I looked at a jar of peanut butter weighing 18 ounces, it contained 2,850 calories. I started using peanut butter as a sweetener to my meals. I stay away from sugar, although I did accept a birthday cake from a friend, which had to have sugar; but that is another story. Anyway, I apply peanut butter to my whole wheat bread.

The calorie count on the 18-ounce jar of peanut butter gave me an idea about what 1,000 calories would actually weigh. By dividing 18 ounces by 2,850 (calories) I got 158.3 calories per ounce. Taking that further, I divided 1,000 (the necessary calories to maintain 100 pounds) by 158.3 and I got 6.31 ounces (which is what I would have to eat to get the 1,000 calories).

**** *TO GAIN WEIGHT IS A PIECE OF CAKE, DON'T DO IT.* ****

Eating well for the single person

This section is more for those readers that live alone, or people that want to eat as well as possible without having to cook every day. There are different items:

To begin with, let me just say that as soon as I started eating better I started to notice my health improving. I continue to eat well; I also exercise every other day, and benefit very much from it. I know that is the way to stay away from doctors digging into my colon, prostate, heart and who knows what else.

At this point I am still looking for a colon doctor with a smaller finger and who will not have such a gleam in his eye as he tells me that he expects me next year, same time same place. Just kidding!

Here is a diet that I put together based on recommended foods:

With this diet I cook fewer times, bigger quantities, and know that when I am ready for my meal it takes five minutes to heat and eat.

*** ***There is no greater virtue then a conquered vice, and there is no greater vice then a conquered virtue.*** ****

I found that eating out alone was boring. I found myself eating what was available on the streets, which was high in fats and sugars. That was what existed as what I had to eat if I wanted to survive. When I was ready to change I started preparing different types of meals, quite varied, to keep from getting bored. It has worked well to get my weight down from 155 pounds to 122 pounds.

Coffee

It is said that too much coffee is bad for us. I have developed a mix: I take one pound of regular mild coffee (fresh ground is better), then I mix it with one pound of garbanzo bean coffee (which is a great coffee substitute).This gives me two pounds which I call half and half. The beauty of it is that even as you continue to consume the same amount of coffee as before you still have reduced your coffee consumption by half. And being that garbanzo beans are high in protein you should get good energy burst from it.

I also prepare another mixture on the side, which involves mixing different dry flavors to spice up my 50-50 mixture of coffee; I use chocolate, cinnamon, nutmeg, ground cloves. And from this mixture, I add a small amount to my 50-50 mixture, so that when the water goes through the coffee pot, it also picks-up the flavors, but it does not stop here, because before I start the coffee pot, I also add a few more flavors, that I have already put together using different liquid mixtures like vanilla, anis, brandy flavoring, coconut, almond extract and orange. The variety of combinations can be endless.

I use and eyedropper to add this liquid mixture to the water while the coffee is brewing. I put other flavors in, whatever I am hankering for at the time. You want the flavorings brewing with the coffee to get that really great blend when it is ready.

Whole-wheat pancakes

This one is easy because all you have to do is follow the instructions on the box, and I use dry milk, but not so fast, what I do is that I also add a little oats, and ground corn meal for extra fiber. For added taste, as the pancakes are cooking on their first side, I add raisins dry granola on top, and I use the biggest frying pan, and this is because after the pancakes cool off I cut them into 4 sections And I also make as many pancakes as I can, for I store them in the refrigerator, So that when I have my coffee I sometimes use one slicc of whole wheat bread with one slice of pancake, which I do in the toaster.

I stopped using sugar in my coffee. At first it was a little difficult, but I did get use to it and I never bother with it anymore. Actually, I used to enjoy honey in my coffee but it is too high in calories when dieting.

While dieting you should remind yourself that it is a temporary situation, your body and mind will find it easier to accept.

I also like to toast one slice of whole wheat bread with half a bagel. When toasted, I add a few drops of mustard for taste. The slice of whole wheat has eighty calories; the half bagel has seventy, so the total is 150 (without the mustard).

There is a saying: "An apple a day keeps the doctor away." I say: "Use whole wheat bread to keep the colon doctor away."

Beans

Later in the day I have one well prepared meal. This meal is based on the fact that we should eat vegetables, fiber, and proteins. Again, for reasons of variety I like to take out time and cook different bean meals.

I picked beans as an important meal because they are high in fiber and protein, and low in fat and cost. I began having meals consisting of just one type of bean; like just black beans or red beans or lentils. Then I got curious about how it would be to mix and match, and that way enjoy a greater variety of flavor and nutrition.

I find that this cooking "in mass" is great because it leaves me plenty of food that I can have ready to eat in five or ten minutes. When friends come over I can offer them something quick and nutritious.

I start by purchasing one pound of every kind of dry bean I can find at the supermarket. By the way, in most cases, dry beans are really cheap. Let me suggest that you try different precooked canned beans, maybe just one kind at a time, to be sure you like them.

OK, here are some of the beans I find in the supermarket: lentil, green and brown, all the different sizes of white beans, all the different red beans, pink beans and black beans, black eye. I usually find about twelve different types of dry beans. I mix all the dry beans together and divide them into smaller packets for dry storage.

When cooking time comes, I use the biggest pressure cooker available for cooking. The bigger size allows me to cook more in one batch, when I do this right I save a lot of trouble and time.

Leave the beans soaking over night, or at least for a few hours before cooking; this makes them softer, easier and quicker to cook.

By the way, to reduce the gas that beans can cause some people to have do not cook them in the same water you use to presoak them. Also, cook the beans without condiments or flavorings and then eliminate that water also before final preparation. You will lose some of the nutrition in this process, but still, it is a good idea if you do not want the gases.

Cook the beans until the largest of them gets soft. I have found that at around fifteen minutes I hear the hiss of the cooker and the beans are ready. I then shut the stove off and let the cooker cool on its own; this takes another fifteen minutes more or less.

I then proceed to add five six-ounce cans of tomato paste and one can of creamed corn. I also add the condiments, to my taste.

For different batches I add different meats. I can make beans taste like chicken, beef, liver, ham, rabbit, goat and any other meats I can find. I also like to make a small batch without meat, for when I feel like a vegetarian and for my vegetarian visitors. One of my favorite batches is the one I make with different kinds of sea food, like tuna, salmon and codfish. Many other varieties can be left to one's taste.

A pot of one my batches will render about fifteen 24-ounce jars. When it is time to eat, I go to the refrigerator, I open one of these jars and I add precooked brown rice, which is part of my main hot meal.

I have accumulated many glass jars that I buy with my applesauce, and I also buy the 20 oz jar that I buy of olives with no pits and filled with red peppers, which I dice into smaller pieces which I then add to the beans not so much for their taste, but for their nutrition. They are excellent for storing meals in the refrigerator; and the meals keep for weeks to months.

Before pouring my cooked beans into the glass jars there are a few things I do to improve cleanliness and freshness. After cleaning the jars, and before pouring the beans in, I put the glass jars in the microwave for 3 minutes; this kills the bacteria that may exist on the glass. I then make sure the beans are completely ready and at their hottest in the pot. That is the time to pour the beans into the jar.

My practice is to then take the lidded jars and run cold water over them to clean the run offs and set them to dry at room temperature. You should hear the lids pop after a few minutes; this is a good thing because it means the jars are airtight. Those that do not pop I separate, these will be eaten first.

For those of you that are interested in why the lid pops down, let me explain. You have applied heat to the beans while cooking them. You pour the beans into the preheated jars. Then you tighten the cap as much as you can.

You will notice that the cap will be a little bowed outward in the middle. Due to the heat the beans are pushing and expanding. The more heat the better, up to a certain point.

As the beans cool down at room temperature they will contract and with the same force that had been used outward they will go inward. The lid comes down along with the contracting beans, tight enough to keep bacteria from even fitting inside. By the way, this makes it virtually impossible for your next batch to have bacteria.

And in the refrigerator cold will further the contraction, this will hold the lid so tightly down that I have had to force lids up with a small spoon as a crowbar. Vacuum packed. The beans can be stored for months.

This is why when you are buying canned food you should check to see that the lid is not up. If it is it means that air has gotten in, bacteria is growing, and this is bad for you.

Next I label the lids: P for plain beans, C for chicken, B for beef (grade A of course), H for ham, S for seafood and R for rabbit. Actually, if I did not label them I would probably forget what is in them half the time. I tend to eat more chicken than any of the others.

Here is how I prepare chicken. I buy ten pounds of cheap chicken and put it in the pressure cooker. I let it cool down overnight. When I get back to it the fat is floating, I take that and after peeling the skin it gets mixed in with the dog food, they love it when I cook.

I remove the bones, which I have to hide from my black and white German Shepherd dogs; we all know they should not fool around with chicken bones. After cleaning the bones and everything else from the chicken it goes into my 32-ounce plastic yogurt containers, which I use for two different servings. For the bean mix. I take what I will be using that day and the rest goes into the freezer. I mark them chicken. From the freezer they go into the refrigerator before being used.

So, on different occasions I make different batches with different flavors.

By the way, aside from using tomato paste and creamed corn along the other ingredients mentioned, I also mix in six eggs.

Macaroni

Macaroni with wheat and oats, I also use olives in macaroni, which along with other pastas, I like very much. I start by buying an assortment of them; I recommend whole wheat pasta above the rest.

Get the biggest pot and pour the water in. Something that works for me is to not add anything to the water (oil nor salt), but you have to stir the pasta continuously.

I make and assortment of pastas and store them in the freezer. At first you should make small amount until you figure out what fits in your freezer. The pastas can be stored in plastic 32-ounce yogurt containers. They get marked MAC on the lid so as not to confuse them with chicken, which I also store in the freezer with the same kind of container. You can also store the macaroni in the 24 ounce preheated glass bottles in the lower part of the refrigerator.

So getting back to cooking, I should mention that I buy different cans of spaghetti sauce that are spiced for spaghetti, some say that they have mushrooms, other say they have cheeses, some say meat flavor, some say plain spaghetti flavor, well I buy all of them this way, I am not missing out on any thing.

I also add fresh tomatoes to these canned assortments. Then I add the real beef meat along with half a dozen beaten eggs, along with some wheat pancake mix and ground corn for fiber. A portion of this gets done with the boneless chicken. All of this goes to the freezer in the plastic 32-ounce yogurt containers, for future use.

Brown rice with barley

Please do not be bored, dear reader, because this is not yet over. You see, I also like to make brown rice, to which I add a small amount of barley for extra fiber to a two pound bag of rice. During cooking I use no condiments, the beans already have them. The rice gets cooked in Corning Ware, in the microwave oven, which is good because you can walk away and do something else as the rice is cooking. So far no burnt rice. In the end it all gets stored into the 32-ounce plastic yogurt containers.

The two pounds of rice with barley will render about 15 small portions, when added to the beans; actually, mixed together just right you get something like a bowl of chili. Anyway, I label my rice containers with a big RI.

Vegetables

This is something like the beans, preparing my vegetable meal. It is a fact that each vegetable has its own nutritional value. This tells me to buy the freshest vegetables available, especially potatoes and green bananas, and whatever needs cooking, but in small amounts. Then I go to the frozen foods area and buy all the different frozen vegetables.

I cook all the vegetables that need cooking in a big pot with water. When they are done I let them cool down. Then I add all of the frozen vegetables and mix them together. I add vinegar to the vegetables that are still in their water (this helps them keep longer in the lower section of the refrigerator). Before putting it in the32-ounce plastic yogurt containers I add dry mustard for added taste, and tofu to what I will presently be eating

Then some of the vegetables go into smaller plastic containers for freezing, I leave some space for the 9% expansion freezing brings. Frozen vegetables can last for months.

Guess what letter the vegetable containers get lettered with!

So that when I decide to eat, I have a wide variety of hot nutritious meals that are just minutes away from eating.

❁~~~~~~~~~~~~~~~~~~~~❁❁~~~~~~~~~~~~~~~~~~~~❁

****** The first thing that is needed to enjoy a good life, is to enjoy good health ******

⌘~~~~~~~~~~~~~~~~~~~~⌘~ ~~~~~~~~~~~~~~~~~~~~⌘

Dessert

Let me say that since I could never stomach milk, and we do need a certain daily amount of calcium, I use yogurt which is a very rich source of vitamins, but more important, yogurt will reproduce itself inside the stomach as very healthy bacteria at stomach temperature. It will keep your stomach very clean.

For those that have never tried yogurt I recommend you start out with a small teaspoon to check out the bitterness. You might have to give it some time to get used to it; it is an acquired taste for some people, but it's well worth the effort.

Here is what I do when it comes to yogurt; as you should know by now, I buy the biggest and cheapest generic brand. To this I add the cheapest brand of the 24-ounce glass jars of chunky applesauce; great jar for storing beans.

I then divide this 32-ounce yogurt in half, which I use for two different days.

I then place this yogurt in the freezer prior to eating. With proper timing it will look and feel like ice cream, sweetened by the applesauce Sometimes I add some of the liquid coffee mixture; there are many ways to flavor what was just plain yogurt.

When dieting I find that staying away from fruits that are high in sugar will help me bring down the weight faster. And fruits have become so expensive that I virtually only eat them at Sizzler. It is highly likely that I am well known at the Sizzler I usually go to, because I am the one who will pile every fruit on display on my plate. Many trips are made by me to the salad bar on one of my visits. Money's worth! Thank you very much!!

I am fortunate in that since I take care of myself I have not yet been told that there are certain foods I should not eat. While dieting I convince my stomach that sooner or later we will get even at the Sizzler salad bar binge. Their sweets are totally destroyed by me, ice cream, apple pie, sweet toppings, but only for that day. Make no mistake about it, after a Sizzler meal it takes longer to drop weight again.

Once I was on a fat free diet, it consisted of eating my bread with 95 to 97 percent fat free cheese, and 95% fat free chicken and turkey ham. You know what? I could not drop weight with it, even as I kept a diary of my daily caloric intake. So I stopped using the 95% ham and cheese deal, which I must say was tasty but not good for losing weight.

Without preaching let me just say, we should not forget or ignore what we eat, for it is our fuel for the moment and it maintains our human existence.

You can imagine that my friends find jars and jars in my refrigerator when they come over. And they do, it is a veritable jar jungle in there!

Tea

As a cold drink let me offer you iced tea. I never drink water directly from the faucet. I use different types of teas and mix them in various ways. Here is one: I have a pitcher which I fill to the top, and to this I add one tea bag of mint and one of cranberry, both are decaf. After a few hours in the refrigerator it is ready. I have found that if left overnight it will taste better; I do not boil the water. Another mix I like is peppermint with strawberry, also decaf. In the end I add concentrated lemon juice. This is what I drink when exercise to take care of the rising temperatures. On days when it is hot, coffee will not do.

I use to question the cost of good food, but I have found that it makes up for it in not having to see the doctor more often and in not having to use more medicine. I gave up certain luxuries, but what the heck. Food costs for me have gone down while the taste, nutrition and quality I get have gone up.

Remember

What we are really doing is looking for a better way to keep from having to see that group of doctors that exist only because you do not take care of yourself. Do we want to pay them to take care of us? I see them once ever few years just to see if there is anything that needs attention. Thank God, so far I have not needed their scalpel service, or pills.

Your body is a gift and it is up to you to keep it clean and in the best operating condition.

So, for those who need a change like the one I made, go to the store and buy the things that will do you good in the long run, for we can only do our best.

*** IT has no FAT areas. Just kidding! ***

3 meals a day

IT found it was necessary for us to have three square meals a day for our primitive body to have the energy needed to keep up with the industrial revolution. Now we need to use less food to exist in this high tech society. Many have already found out that we now need to be as lean as possible and give the fat to the robots that are continuing to run with what still exists as the industrial revolution; and see if they will get fat too.

For mankind, fat was a form of stored energy with which we survived the lean, difficult times. As for the robots, they can use this fat as a last resort from locking up from friction (from working so hard), as a substitute for grease. We should also remember that from that bulky industry we now have the industry of electronics that ***IT*** reshaped into, which is run by ***IT*** more as robots and less as humans. We should be grateful to ***IT*** for the electronics, computers and circuits that are necessary for the space ships that will take humans into ***ITS*** other area, that we call D*ark Matter,* or ***ITS*** cold, clear nothingness, outer space.

Neck and back pains

The dry swim: In a standing position pretend that you are swimming. Raise your right hand over the left side of your head and bring it down like a swim stroke. Repeat, but with your left hand. As you are doing this move your face from side to side, this will loosen up the neck. Also, as you raise hands over your head, move your hips to the left and to the right, this will loosen your middle. Try to get to the point where you can repeat this exercise one hundred times, we don't want bones welding themselves together due to all that sitting.

Pull-ups

Pull-ups are a very good exercise, very efficient and inexpensive. A pull-up bar can cost $10.00 or less and is easy to install. With pull-ups you use your own body weight to stretch the muscles of your shoulders, elbows, wrists and fingers. Even your whole back and waist will feel the tug of gravity as you hang and pull yourself up. And you will also feel your heart pumping faster

I use a pad to jot down what I do when I exercise, it is interesting to go back and see one's progress documented. When I started doing pull-ups I barely did two, little by little I kept building up until I got to 15 times , I then take a few seconds to catch my breath , and I then do an other 15 pull ups , and I do this for a total 7 times the 15 pull ups. That is where I leveled off and stayed.

To these same pull-ups I added an exercise to strengthen my abdominal muscles. As I was hanging and pulling myself up I started raising my feet in front of me as slowly as possible, without bending my knees.

In a way, it was like getting two for one. Because of the extra effort it took to add this routine to the pull-ups, there was a drastic reduction in my repetitions, but little by little I got it back up to 105 again. Raising ones legs while not bending the knees really strengthens the abdominal area.

⌘~~~~~~~~~~~~~~⌘⌘~~~~~~~~~~~~~~⌘~~~~~~~~~~~~~⌘

*** Laughing is a good exercise. It's like jogging on the inside.***

✿~~~~~~~~~~~~~~~✿✿~~~~~~~~~~~~~~✿~~~~~~~~~~~~~~✿

Alzheimer's Disease

I have a plan for detecting my own potential onset of Alzheimer's. When I exercise my body, I keep my mind busy by counting the repetitions. I count 100push-ups and 100 pull-ups. I count to 100 for bench presses and the same again for leg exercises. When I can no longer count these hundreds of repetitions, I know that Alzheimer's is just around the corner. Today I am still grateful for being able to exercise and think at the same time and am hoping there is truth in the saying "use it or lose it."

Since in most cases the mind cannot be stopped from thinking, we have a few options from which to pick:

- Do Nothing—allow the environment to manipulate the mind.
- Daydream—manipulate your own mind.
- Exercise—Think about what ***IT*** is and how ***IT*** operates.

❁~~~~~~~~~~~~~~~~~~~~~~~~~❁❁~~~~~~~~~~~~~~~~~~~~~~~❁

IT reshaped into doctors and the necessary medicine for us to exist longer.

⌘~~~~~~~~~~~~~~~⌘~~~~~~~~~~~~~~⌘~~~~~~~~~~~~~⌘

Arthritis

As for arthritis, I believe that by using exercise to work through my pain, I am eliminating the pain and preventing future pain.

A softer shave

I have traveled before to find that I had forgotten to pack shaving cream. So instead of using the hotel soap, which would be too harsh on the skin, drying it rather than moistening in preparation for shaving, I used the hotel cream rinse for hair. I learned that in addition to being cheaper than shaving cream, cream rinse left my skin feeling even softer.

For cleaner and healthier teeth

My formula for cleaning the teeth is as follows: Mix large quantities of baking soda, which helps whiten and remove stains, with plain salt, which helps disinfect. Dab a small amount in the hand, add a few drops of peroxide to help fight germs, and brush. I then follow the brushing by gargling with peroxide and rinsing with plain water.

⌘~~~~~~~~~~~~~~~~~~~~~⌘~ ~~~~~~~~~~~~~~~~~~~~⌘

**** ***Thank you God for taking care of me physically, in the doctors you exist as. I am grateful to be able to physically see more of you as you reshape in this existing moment.*** *****

PART # 2

If nothing existed

I would like to start this section by asking you to look around your self and notice all the things you see: your refrigerator, your radio, your TV, buildings, cars, people, trees, birds. All are made of matter. We easily take these material things and beings for granted because we believe everything that exists should exist. But these things only exist because ***IT*** exists. God could have existed in a lifeless universe. Imagine if all the things we see and do were not possible because there was no God, no pure energy. Imagine that there is no Universe, that there is nothing. With nothing, nothing happens; you would not even be here to think and observe your surroundings, obviously. We can be grateful that whatever this God is, ***IT*** does exist.

The word "God" could use some revising, however. The dictionary's definition of "God" is this: One who causes to come into existence; a person who grows, makes, or invents things. And there is nothing wrong with this meaning, but it does mislead with the implication that something can be made from nothing. Let's examine an occasion about which one might say, "I have created." Take for example a painting that I "cause to come into existence," for it hypothetically did not exist before the occasion of my creating it. Before beginning to paint, I would have to find the materials necessary for painting: canvas, paints, and brushes. In other words, in order for me to create, I must use materials that I did not create, materials that have preexisting atomic structures. Additionally, that raw material that is needed has to be found, and the fact that it is "raw" implies such material must be independent of the one creating. If God is "the One", however, nothing could be independent of ***IT***.

Accordingly, to create the Universe, God would have needed preexisting substance: stardust perhaps? No, not stardust, because before the Big Bang, there were no stars! There was only ***IT*** with all ***ITS*** weight concentrated into one singular point. This is why I prefer to call God ***IT*** and why I use the words reshape, mold, transform, and rearrange to describe the way things come into existence. "Creating" does not signify the oneness of ***IT*** as all the pure energy that has always existed, whereas "reshaping" does. Reshaping entails raw materials that have always existed even before the Big Bang. The human mind simply prefers the term "created" because the mind seeks recognition as the most important thing that exists, whereas in the case of my painting, it is the result of ***IT*** reshaping first into atoms and then into the materials that I require before I can even begin what I would call "my creation". And now comes the other part: in order for me to create anything, I first have to exist.

Using the word create thus tends to inflame and confuse me as a human mind, for anything outside of me is there only because ***IT*** reshaped into it. Still, our minds want to believe that because we sexually reproduce, we create our own human kind. But we do not create. We merely participate as humans in the reshaping of our God's Universe.

❁~~~~~~~~~~❁~~~~~~~~~~❁~~~~~~~~~~❁~~~~~~~~~❁

** ***WE ARE NOT CREATING; WE ARE RESHAPING*****

⌘~~~~~~~~~~~~⌘~~~~~~~~~~~⌘

A show called omnipresent

Many human players understand omnipresence as a manuscript called the perfect plan.

This manuscript is our God, pure energy, and it's not billions or trillions of years old. Billions of years ago, the human world did not exist. When the human actor arrived upon the stage of planet Earth, there were already many other living actors, larger and stronger actors. The human actors had to use everything within their reach, even weapons, to protect themselves. God realized that in order for humans to exist, the larger and stronger animals had to be eliminated and were thus reshaped into something more functional within their current environment. The perfect plan is thus a perfect show, one that started many of Earth's rotations ago. The performance has continued until this current moment and will continue into what we call the future. We are all actors from the beginning of our existence, in ***ITSELF as ITS*** show.

Somewhere in our evolution, human actors began to believe that they were the most important part of this show called life, but we must remember: ***IT*** resides within each of us. You are ***IT***; ***IT*** is you. Everything that exists is ***IT***. ***IT*** will provide for you what you need. Do not assume that because you have a mind to think freely and spontaneously that ***IT*** is not in control. Many people who play authority roles become confused as to how much control they actually have in this play. These authorities fail to realize that their roles will end. ***IT* is the only producer and director and the one thing that has existed from the beginning that will continue to exist** until *IT* decides the show must end.

As the producer, director, and the actors of this perfect show, ***IT*** makes ***ITSELF*** available to assist other parts of ***ITSELF*** that may be lost or in danger and are crying out for directions. God answers in a whisper to provide this simple message: "**Do that which only you know how to do in my play, and do it only at the moment that you are supposed to. And don't ask why.**" The director will make the necessary adjustments to reshape you into your next role. If your role is an unpleasant one, be grateful that you are still a player on this stage. If the role that you receive causes you pain, ask for a new, more pleasant role; you may be provided one. You must be open to accepting the instructions that you receive. Some have received messages from perfect masters explaining the show as God and the pure energy in ***ITS*** omnipresence, but only God knows what parts are important and required in this **perfectly reshaping play**. Humans were preprogrammed to be servile. Servility makes it easier for us to exist, for God is orchestrating everything in this show. ***IT*** is the director, producer, and actors all at the same moment.

IT underwent much extinction and reshaping before arriving at a human intelligence capable of reading, writing, analyzing, and understanding where we came from and what put us here. Our ancestors did not know how many of their kind existed or even where on this planet they were located, but now it's clear that we have always been right here inside of ***IT***.

In the time since your birth, you have grown bigger, possibly fatter and ideally wiser, but you have never ever been anywhere else other than this place of omnipresence, which has always existed, even before the Big Bang. You may have moved your housing and you may have traveled to some other location on this Earth, but you have always existed inside of ***IT*** as omnipresent, as the here and now. For this we should be grateful. Without ***IT***, none of us would be here.

And if ***IT*** reshapes so that there no longer exists a need for humans, we would transform into something else. It is possible and probable that a meteor will hit this planet and do away with the human race. Take heart though; we will continue to exist as this moment, for we are part of ***ITS*** pure energy and always have been. We cannot exist outside of this omnipresence. We are made of solar dust and in ***ITS*** own image: we are this God as ***IT*** reshaped.

Even God is made in ***ITS*** own image, for ***IT*** has to continuously be engaged in the transferring of energy, which is to say, ***IT*** does not rest, not even for a second.

IT as life

You are here because you have something called life. I believe that life existed before the Big Bang, and as we can see, it still exists. As our planet was formed, ***IT*** allowed for life to develop into what we are, as ***ITS*** reshaping of that which we know as life.

Your life is housed within a body that started at conception, developed into a baby, and kept reshaping into maturity. Life begins with atoms. The male sperm and female egg are made of atoms. The sperm and egg combine so that they, as a duality, can reshape into life. The human body becomes the housing for life.

The life force that exists as you does not change or reshape throughout your life; ***IT*** is constant. I find it strange that life is the only thing that does not change.

This also means that the part of ***IT*** that we understand as life is not subject to change. Let's look at the life of an ant. An ant is very small. An ant has life, the same type of life that exists within us. Life is the same for all but varies depending on the housing.

The evolution of the human form has allowed us to use information that is available at any given moment. The ant does not have the same degree of this ability. Human beings are able to participate in the reshaping of things differently than ants. Different forms of life have different tasks, functions, and purposes.

Change happens as a result of a positive and negative energy coming together. This is similar to when a male and female make physical contact. This causes something new to come to life, such as a baby, sometimes.

~~~~~~

We are the result of a constantly changing body in the presence of a constant that we call life.

⌘~~~~~~~~~~~~~~~~~~~~~~~~~~~~~~⌘

Life as we know it has a range that encompasses our deepest waters, all earthly terrain, and large portions of our skies. Events and beings are the products of ***IT*** reshaping ***ITSELF*** within ***ITSELF***, and being made up of pure energy, they exist in the same moment that we are now, in the place of omnipresence. Our history and present life within this omnipresence is 100% ***IT***, and being of this life, so are we.

In our modern language, we have used the word life in so many contexts, and even to refer to non-living things, that it's evident we don't altogether understand the meaning of life. When seeing a freshly washed car, someone might exclaim, "It has come alive!" When we admire a lifelike painting, we actually question whether it is alive. Or about a house, we might say, "look how alive it is when the lights are on" or "see how alive the room feels now that it's redecorated." There are more ways, too, that we use words related to life. We speak of the life of a star, planet, or galaxy, for example, and when we do this, we should be aware, that we are saying that these celestial bodies exist with mobility.

⌘~~~~~~~~~~~~~~~~~~~~⌘~ ~~~~~~~~~~~~~~~~~~~~⌘

The presence of all the extreme political parties is ITS way of finding all the possibilities that IT can exist as.

~~~~~~~~~~~~~~~~~~~~~~~~~~~~~~~~~

Life is ***IT***, and only ***IT*** has mobility in ***ITS*** way of existing as a constant life that can exist as millions of bacteria so small that they can exist with mobility upon the head of a pin or as the human mind that enables me to write this book and you to read it.

Here is something to think about: Since ***IT*** is one, and everything is happening inside of ***ITSELF***, what we understand as life cannot exist if this quality were not there to begin with as ***ITSELF.***

So that if ***IT*** exists as LIFE, then even a rock or a steel rebar has this quality inside of it as ***ITSELF***, even if these objects do not have mobility, for the rock and the steel rebar have both ***ITS*** dual ways of existing, and this also applies to everything else that exists. In the same way, everything that exists also has all the other qualities that ***IT*** has.

Life is where IT exists with mobility

The need for mobility is why life may have started in the ocean. ***IT,*** as a water-warehouse of natural resources could provide ***ITSELF*** the elements that would give ***IT*** mobility, most importantly the number one element, hydrogen, which has the capability of forming bonds. Without hydrogen we could not exist. Hydrogen's bonding power provides ***ITSELF*** a way to reshape without having to repeatedly return to the formation of atoms, the first step of ***ITS*** reshaping. By bonding hydrogen atoms, ***IT*** reshaped into water and the smallest form of ocean life that could exist with mobility.

Because water can absorb great amounts of heat, by forming life first in water, ***IT*** could also protect itself as ***IT*** began to form the first eyes, ears, and all the other organs of the newly mobile life forms. Under this protection, ***IT*** could explore more possibilities of existence. There are many volumes of books on the life forms that exist in water, on land, and in the air.

As we observe all that is alive, we can see that ***IT*** also used atoms to form the components of DNA and RNA, which keeps ***IT*** from having to return to base one when reshaping ***ITSELF*** as life with mobility. Ocean creatures went from single cells to having brains, eyes, and mouths. ***IT*** has evolved through millions, billions, trillions, and likely googols (10^{100}) of life forms as ***IT*** continued to reshape with mobility, and in this present moment where we now exist, ***IT*** is still mobilizing life. But when ***IT*** began to roam as life on land, there was still no sight of you or me.

This is also why we know that life can exist in areas of very high temperatures because it is not the temperatures that determine life; as long as there is water, ***IT*** will be searching to be present as life with mobility. And I doubt this is the first time that ***IT*** has reshaped with mobility, for ***IT*** could have done this somewhere else in ***ITSELF*** as this Universe. I say this because if ***IT*** is life, then ***IT*** knows what ***IT*** can reshape into as ***ITSELF*** as life.

IT reshaped into water so that ***IT*** could become a cell, so that ***IT*** could reshape into tissue and continue to change into organs. Cells, tissues, organs: they each have mass, yet they each continue to be ***IT***, and looking to the smallest form of what is known as having life attached to it, we should remember that this life form is ***IT*** in totality. And remember as well that life is not made of matter; it is made from ***ITS*** constant nothingness.

I feel that most likely ***IT*** is a wave or vibration that started before the Big Bang and that afterwards reshaped into atoms, which behave as both waves and particles that vibrate. And as atoms, ***IT*** reshaped into everything prior to the beginning of living beings. Furthermore, in ***ITS*** continuous reshaping into all existing possibilities, ***IT*** arrived at what we call human life, which has the most intelligent brain that exists on this planet. ***IT*** reshaped into a mind that needed a body and skeleton to house the heart and all the organs necessary for ***IT*** to exist as humans. Life is nothing other than ***IT,*** as one, existing in all places as the same moment known as omnipresent. All the billions of life forms that exist are just one as ***ITSELF*** with mobility. ***IT*** is in all that we see alive.

Life ***ITSELF*** has intelligence. The human form allows for the human brain. The human brain contains what we call the mind.

As ***IT*** reshaped ***ITSELF*** into a planet that we call Earth, atoms further reshaped to produce water. Water made life possible. Atoms reshaped into cells, which reshaped into tissues, which reshaped into organs, which reshaped into organisms, which were the first forms of life on this planet.

I see this universal reshaping as a circular action, for it's typical to perceive the Universe as circular. The planets, stars, and moons are round. Many living cells are round. Atoms are round. What is known in meditation as the third eye is round.

It is a law within ***ITSELF*** that ***IT***, as a total, will not remain in one shape. To do so would defeat the process of reshaping, of changing. ***IT*** incorporated this law before ***IT*** became the Big Bang.

Life as 21 grams

It is said that when a person dies his body weight drops by 21 grams. Now this may be true, but if we take the smallest life form that may exist and weighed it, this life form would weigh less than these 21 grams that are associated with the human form.

Life is ***IT*** with mobility, and this mobile life does not have weight, because in having weight, life would then be made of matter and thus be in constant change. And as I have mentioned, life is ***IT*** as a constant. So I pass this problem on to the scholars who can look deeper into this and would like to share their information with the rest of us, for if we can find the smallest form of life that exists and deduce its mass in terms of weight, we would get closer to finding the weight of life as ***ITSELF***.

One becomes millions

Here is an experiment for the pros on this subject called life. The experiment consists of putting a living organism in a closed environment such as a one-meter box or airtight container, ideally one that is fire resistant and has a window. In this air-tight, sealed box, put a controlled, living organism, which would have water or a water-like fluid as its cellular base.

The next step is to apply heat to the box so that the life within dies. If the box is secure, nothing should be able to enter or escape it. Having a totally sealed container will maintain the conservation of organic matter inside it. After applying the heat, the box would then be void of life activity; the organism would be destroyed.

The inquiry here is to see whether or not life can regenerate with what existed before.

This experiment is not humane, so I'm not asking you to do it, just recall a time when you have seen a recently deceased animal in the wild or on the roadside. You know it had just one life when it was alive, but as it decomposes, it becomes millions and millions of other life forms for as long as water is present in the dead animal's tissues. It's not an attractive scene to imagine, but it demonstrates that life is ***IT*** where ***IT*** exists with mobility. Be it here on Earth or in any other part of ***ITSELF*** known as the Universe, life exists wherever the conditions permit water to be present.

But now that we know we are here and each of us has a mind, a heart, and all the other organs that exist as our physical form, where exactly is life located?

Life is not within our bodies the same way an organ is. To see life and become closer to ***IT*** as life, we would need to engage in meditation, for life as ***IT*** exists as color, sound, and motion. And to reshape into life with motion, ***IT*** first had to convert from being a fixed dense energy to an energy with mobility.

GOD

The human mind has attached various meanings to the word God. We should remember that there was a moment when we did not exist, when our planet did not exist, but ***IT*** did exist, and ***IT*** did so without the need for this word, "God".

Yet we refer to God as a pilot in control of our lives. And certainly, when on a plane in the sky, where we are without control as to what could happen to us as life and we cannot walk away from our situation, such as when we encounter turbulence, we will thank God for keeping us safe. Yet when we get as close as possible to having our feet back on the ground, where we again have the feeling of being in control, we lose our need and gratitude for the pilot that we called God.

We also use sayings such as "he thinks he is God" or a similar one: "he is playing God." I have not heard these expressions in reference to a female, as in "she thinks she is God." Odd, is it not? Perhaps this is because in our culture it has been predominantly men at the controls. Surely as women gain their share of these controls, we will be hearing new sayings.

In my travels through life, I was once told that God is an acronym: **G.O.D. G** stands for ***IT*** as the Generator, and I can still see this to be so. **O stands** for ***IT*** being the Operator, and I also agree with this, and **D** stands for Destroyer, and here is where I disagree. ***IT*** is not destroying; ***IT*** is reshaping. This is why I still prefer to use the word ***IT***.

We have attached words to God that imply there is more than one. Ancient Greeks or Romans would say, "In the name of the gods!" But the perspective of multiple Gods misses the meaning of omnipresence.

There are two gods, however: there is the god of the mind, a righteous God that will see things that are wrong, such as negligence, hate, ignorance, and war; and, there is the God that is running the Universe. We seldom remember that when we discuss God, we must associate ***IT*** with everything that exists as omnipresent.

Many wars have been fought over which God is the true one. I believe there is no other word besides omnipresent that unites everything that we see and do not see. God (pure energy), this omnipresence, is in all places at the same moment. This is why I have chosen to call God ***IT***.

The meaning of ***IT*** becomes clearer when we hear a genius's perspective. For example, Albert Einstein said, "God does not play dice." For whatever ***IT*** was to Einstein, ***IT*** was not playing a game, which is to say that unlike a game having a loser and a winner, ***IT*** has no opponents. Every thing is ***IT.***

As for me, I do not attach words to ***IT*** besides ***IT***, for I know that ***IT*** exists as me and as every thing outside me. I do however use words to thank ***IT*** for letting me be here and for giving me this moment where I can think more about ***IT***, for ***IT*** could, if ***IT*** wanted, have me laboring with a pick and shovel, or worse, have me on a sick bed, feeling pain before I leave this existing moment. I know the pains I now have are to let me know that there are a few things that I have to do in order to keep what I already have in the best condition possible.

I do also say to ***IT***, "You are incredible—in size and in all that you are doing, as yourself." I also ask ***IT***, for I know ***IT*** is always with me as omnipresent, to help me never again lose my focus on ***IT.***

Many people focus on ***IT*** or God by way of symbols: a cross, a statue, an icon, or pictures of the Universe. Humans have philosophized and theorized about the meanings of these symbols in relationship to God, yet we should remember that we have only recently become aware of this Universe that we live in. During the primitive stages of human existence our minds were not as analytical; there was no need for it. There was no need for an education as we now know it.

Now that we have the capacity for analysis, we are able to examine ***IT*** as the atom. Looking into the atom permits us to understand more about pure energy. There are professionals in the scientific community that understand pure energy as ***IT***. They believe this because they are aware of energy, and energy is quantifiable. And, since we are made from this pure energy, it would be reasonable to accept that we are this pure energy: this substance of God that we are learning to understand as ***IT*** is reshaping ***ITSELF*** in omnipresence.

❁~~~~~~~~~~~~~~~~~~~❁ ❁~~~~~~~~~~~~~~~~~~~❁

****___Since there is a law that nothing is really created or destroyed, then you will understand that life is something that was not created or destroyed either, as pure energy, or as IT.___****

⌘~~~~~~~~~~~~~~~~~~~~~~~~~~~~⌘

Our inability to make the connection between God and pure energy before now is due in part to our ability to function without the need to know. It is also due to our lack of technological training. As we advance in our technical ability, we will come to understand ***IT*** even more. Currently, more information is becoming available because more and more people are joining technological fields as a source of employment. The increase in employment opportunities increases the amount of education committed to the subject.

What we can understand at this moment is that ***IT*** will be using this high-tech society to move from one part of ***ITSELF*** to another. ***IT*** is aware that this planet has a finite amount of time in which to exist as we know it. Through the technology we are developing that will take us into outer space, ***IT*** is reshaping ***ITS*** possibilities to reshape into a part of ***ITSELF*** other than planet Earth. Since this ultimate reshaping is still going to be this same existing moment as omnipresent, many rotations away, there will be constant reshaping in preparation for what ***IT*** develops into.

As one moment of my existence, I too saw God as being that which was on a cross, and I too saw science as being elements that could not be created or destroyed, but I now understand that these elements listed in the Periodic Table can be seen as the planets, stars and every thing that exists out there as this Universe, and that these same elements can be taken and reshaped in to an infinite arrangement of things. Like the chair that we use to sit down in and eat our food or work, the same elements reshaped will be found in my human body.

I too once understood God as one thing and pure energy as another, as something scientific that made my telephone and all the existing materials that exist outside of me as objects. But now with this better understanding of the meaning of omnipresent, I see the cross as a symbol made of elements of pure energy, which cannot be created or destroyed but is reshaping in this place called omnipresent, for everything that exists is a fraction to the minus side of the number line of ***IT*** as one. And this one as 100% does not need a name; ***IT*** is beyond that. ***IT*** is even beyond time. There is no rush for ***IT*** to reach a certain point at a certain moment. ***IT*** is all that exists at the same moment. ***IT*** is beyond the abilities of the human mind to conceive what ***IT*** will reshape into. ***ITS*** possibilities are infinite.

We can understand what ***IT*** was before ***IT*** reshaped ***ITSELF*** into this moment as our lives. We cannot see what ***IT*** will reshape into as what we call the future. When the mind thinks about what it is going to do tomorrow, it has to remember that it will have to hold off until the planet rotates. I have accepted that all of what I am will have to keep changing until the moment that I will no longer exist as this gift called life.

❁~~~~~~~~~~~~~~~~~~~~❁ ❁~~~~~~~~~~~~~~~~~~~~❁

*****We can call God anything we desire as long as we exist. *****

⌘~~~~~~~~~~~~~~~~~~~~~~~~~~~~~⌘

There is only one master. And that is ***IT***, for only ***IT*** knows what it needs to reshape into. ***IT*** is the master of reshaping for ***IT*** is reshaping ***ITSELF***. Yet, there is no perfect plan; we have no idea what ***IT*** is going to reshape into next. Even as you read this, God is reshaping. I trust in ***ITS*** reshaping, yet to me ***IT*** remains mind-boggling. If ***IT*** is everything including us, where does ***IT*** exist as a place? ***IT*** runs this whole Universe, from that last atom at the end of this Universe all the way to the atoms that make my existence possible. Way before I arrived here, or my parents and grandparents arrived, when we began to exist as humans, ***IT*** has been here as ***ITSELF***, and whatever ***IT*** wants to do with us as ***ITSELF***, as ***IT*** prepares to take us off this planet, to some other part of ***ITSELF***, we really have no say, for ***IT*** is the only one that knows what is out there as ***ITSELF***, and as for me, as this moment, all I have to do is enjoy as much as I can as ***IT*** keeps reshaping this moment called life. Even if what I see looks like madness, I have to remember that ***IT*** knows what ***IT*** is doing and always has, be it here on Earth or out there as this Universe.

❀~~~~~~~~~~~~~~~~ ❀❀~~~~~~~~~~~~~~~~ ❀

***** God recognizes no flag and favors none *****

⌘~~~~~~~~~~~~~~~~~~~~~~~~~~~~~~~⌘

IT is alive

Since everything is ***IT***, life too has to be ***IT***. ***IT*** produced life because ***IT*** is alive. I did not always see my God or pure energy as being alive, but looking closely, we will notice that we are made in ***ITS*** own image and we are alive. What's more, we consume ***ITS*** pure energy: ***IT*** reshaped as food, air, meat, and vegetables, so that we can live within omnipresence.

We describe life as an energy force because we cannot visually see life. Many in fact call the life within us our "spirit" because life is potent enough to move our human body and our spirit is considered that which continues living after our bodies die. Still others say life is found in our soul. While I have no intention of giving you an opinion as to what to think about the idea of a soul, it has been said that SOUL is an acronym for "Source Of Universal Languish." I have been fortunate to learn from ***IT*** of a place where I will not be affected by languishing or what some might see as the grinding wheels of life, where no matter how we position ourselves, we feel as though we are being ground between gears like a grain of wheat in a mill. For in all fairness, the grinding wheels are just doing their part in ***ITS*** reshaping. Therefore, instead of struggling between the wheels, be it the right wheel or the left, I have learned to surrender to ***IT*** and see myself as the pivot, that part of a wheel needed for turning, but away from the grinding of change as ***IT*** continues reshaping.

One of the most precious gifts that I have is my ability to communicate directly with ***IT*** and to seek help from ***IT***, and you can too. ***IT*** communicates with us every time ***IT*** reshapes, which is constantly. Some of us have at times met a perfect master or guru to show us how to connect with the spirit or pure energy that exists within us, and in this way we have come into contact with ***IT***. Sometimes our teachers appear as we pursue our interests. I was lucky to have chosen water as my career interest, for my research on water put me in contact with other related areas and provided me insight into how ***IT*** functions as life with mobility.

Another way I connected to ***IT*** was by focusing on the word "omnipresent." This word led me to understand God as pure energy and everything that exists. I am certain that as more people look into omnipresence, they will better understand this pure energy as God and also as themselves. Many have yet to connect to this internal feeling of pure, universal energy in which we are all living and of which we are a part. Those of you who are interested in making this connection with ***IT*** and would like to do so for free, for ***IT*** cannot be bought or sold; you can contact **Maharaji at (<http://maharaji.org**>),.

As we age, something new always comes, even if what's new is pain.

Ask and you shall receive

Before I could see what was happening to me, I fell into a hole called addiction. By the time that I realized what had happened I was in too deep to find my way out. I looked to the outside for help but found that people could not offer me a way out of my situation.

I asked God, from the bottom of my heart, to help me. I confessed that I was lost, and I did not know my way out. It was at this moment that I became aware that God does listen.

IT can see, hear, and feel because ***IT*** created the human body.

A few days after asking ***IT*** for help, a big change started to take place. This change has lasted to the present moment. The change was noticeable right at the beginning, but I did not understand it. After a few years had passed, I looked back at the many changes that took place and began to understand ***IT*** as I do at this moment.

Life got better and better for me. My friends now ask me how I am doing, and the first thing I tell them is that I am alive, like they are, and then I say, "I eat well, I sleep well, and I feel well; these are the best of times for me; my cup is overflowing." And for this, every moment that I get a chance, I thank ***IT*** for allowing me to live. I see now that since I asked for change, I have received change.

I also asked ***IT*** if it would allow me to know and understand ***IT*** more. ***IT*** began to show me more of what ***IT*** is and how ***IT*** operates. I again have to say "thank you." I can now understand the saying: "Be careful for what you ask because you just may get it." I still talk to ***IT***. ***IT*** is the only one who has been with me since the moment I became alive. ***IT*** is the only one that will be with me when I take my last breath of life. ***IT*** is my best friend.

All existing possibilities

Here is one more way to open your mind to how ***IT*** exists: Imagine that you are God, and that you are everywhere at the same time, and that everything that exists is you, as one, as in you being in all places at the same time. It will help if you remember that you exist as one total heated weight, that exists within the way you exist as one huge, cold, clear nothingness, as your outer way of existing, as a place where you keep your heated weight, that can exist as just one total weight, (like before the moment of the Big Bang) that can be fragmented into very tiny portions (protons, neutrons, and electrons). As you imagine this, remember that you are GOD, as a portion of ***ITS*** weight, that exists within a portion of ***ITS*** total nothingness.

Whatever this visualization may lead to, as to what would you see and comprehend, and if you find something that you would like to share with the rest of us, please send it, so we can use it in understanding ***IT*** better.

Here is yet another way to see this: If GOD is in all places at the same moment, this means that at my moment of existing, ***IT*** is here also, and if GOD is all knowing, then ***IT*** is also inside of me as this existing moment, firstly, as omnipresent, and secondly, since ***IT***, to use the phrase, created everything, ***IT*** is also you and I and everything that may exist from ***ITS*** creation.

❀~~~~~~~~~~~~~❀~~~~~~~~❀~~~~~~~~~~~~❀~~~~~~~~~❀

***** When we speak about nature working through random possibilities, this is the same as IT searching for all existing possibilities *****

⌘~~~~~~~~~~~~~~~⌘~~~~~~~~~~~~~~~~⌘~~~~~~~~~~~~~~⌘

IT as I-S-F-A-E-P

As we look at all that is happening in and around us, we can see everything as being done by ***IT***. I tell myself that ***IT*** is acting as I-S-F-A-E-P (***IT*** Searching for All Existing Possibilities). When we see things that we do not like, let us remember that all that is happening is ***IT*** as ***IT*** reshapes. I can only recommend what I have done, which is to see what comes from ***ITS*** reshaping not as something meant to satisfy my liking, but rather as something intended to make something different happen as ***ITSELF***. With this perspective, I no longer feel the pain and discomfort that comes from not liking what I am seeing. I understand that everything that is happening as events is ***IT*** reshaping ***ITS*** weight into something new to continue ***ITS*** search for all existing possibilities.

❀~~~~~~❀

****Our protesting is one more way for IT to search for other possibilities.***

⌘~~~~~~~~~~~~~~⌘⌘~~~~~~~~~~~~~~~⌘⌘~~~~~~~~~~~~~⌘

What I do is keep an eye on ***ITS*** new ways of reshaping, trying only to see what might come from ***IT*** next, even if what I am seeing is what I once considered a negative action. This is ***ITS*** show. As humans, we have the chance to exist as ***ITS*** life form, in this stage of ***ITS*** existence.

❀~~~~~~~~~~~~~~~~~~~~❀❀~~~~~~~~~~~~~~~~~~~~❀

******When you see a human reach a new world record, you are seeing IT as IT searches for all existing possibilities.******

⌘~~~~~~~~~~~~~~~⌘⌘~~~~~~~~~~~~~~~⌘~~~~~~~~~~~~~~⌘

A word for IT

I invite any readers to join my search for another word for ***IT*** that would denote ***IT*** as one entity operating as two forces as the same moment. I have played with this objective of finding a word as though I am arranging squares from a Scrabble game, for such a word should stem from the meanings if not the first letters of the following words:

Omni, for whatever ***IT*** is, it's everywhere.

Speed, for ***ITS*** mobility and ability to transfer light.

Light where ***IT*** exists behaving as a duality: as waves and as particles.

Empty, for ***ITS*** 99.99% nothingness —plenty of room to reshape.

Weight, for ***ITS*** mass.

Temperature, Color, and Sound, for ***ITS*** heat and cold, visibility, and the whispers of wind and cracks of thunder.

Reshaping, for ***ITS*** function of creating.

Transforming, for doing it all infinitely and perpetually within ***ITSELF***-- ***IT*** is perpetual, which according to the dictionary means never-ending, which echoes ***ITS*** infinite possibilities, and to continuously repeat; ***IT*** is repetitious in ***ITS*** reshaping.

And here's a new one: Magnetic, for ***IT*** functions as two equal forces acting as one, like a magnet—negative repelling negative, positive repelling positive; hence the atom's neutron enabling positively charged protons to coexist and for ***IT*** to reshape into matter. Furthermore, ***IT*** operates as an electrical, magnetic field to produce these changes or moving events.

However, it is not as important to know the definition of these words as it is to know and understand what ***IT*** is and how ***IT*** operates, since the terms God, pure energy, Creator, or any other respective title, in addition to ***IT***, relates to the same power that exists as ***IT***. I can move letters around all day to find a word, and please write if you can help, but I for now have to put this word-search aside, so that I might continue this book.

***** The more you know about IT, the closer you get to IT and the closer you get to IT the more you have to surrender to IT, and the more you surrender to IT the more you become IT, but not as who we think we are, for you will realize that you are a fragment of ITS total weight and that you exist within ITS nothingness. ******

⌘~~~~~~~~~~~~~~~⌘⌘~~~~~~~~⌘⌘~~~~~~~~~~~~~~⌘

An unbalanced universe

The balance in ***ITS*** universe is strange considering that less than 5% is matter and at least 95% is empty space; 10% of things are uncertain while 90% of things generally go the way we expect them to; 45% of the people on earth disagree with the other 45% of the people on earth, and the remaining percent are unsure.

The chemistry of the human body is unbalanced as well. Men have 80% testosterone and 20% estrogen. Women have 20% testosterone and 80% estrogen. This is true for 80% of the population while the other 20% do not have the same make-up.

This could explain homosexuality. Perhaps an expert on the subject knows whether the ratio of testosterone and estrogen is different in homosexuals. And here's another unbalanced ratio; some say the homosexual community is 10% of the population while near 90% is heterosexual, a small percent being neither. Let us remember that ***IT*** is the 10%, the 90%, and all other percentages. What is important is that ***IT*** will always be searching for all existing possibilities as ***ITSELF*** as life.

❀~~~~~~~~~~~~~~❀~~~~~~~~~~~❀~~~~~~~~~~~~~~~~~~~~❀

Memories are what IT uses to remember what IT reshaped from, or how ITS weight used to be or how ITS weight once existed.

⌘~~~~~~~~~~~~~~~~⌘⌘~~~~~~~~~~~~~~~⌘~~~~~~~~~~~~~⌘

The best worker

The best worker is ***IT*** because ***IT*** is unendingly transferring energy. Nobody can outwork ***IT***; ***IT*** never stops, not even for a coffee break. No one can outperform ***IT***; nothing can transfer as much energy as ***IT***.

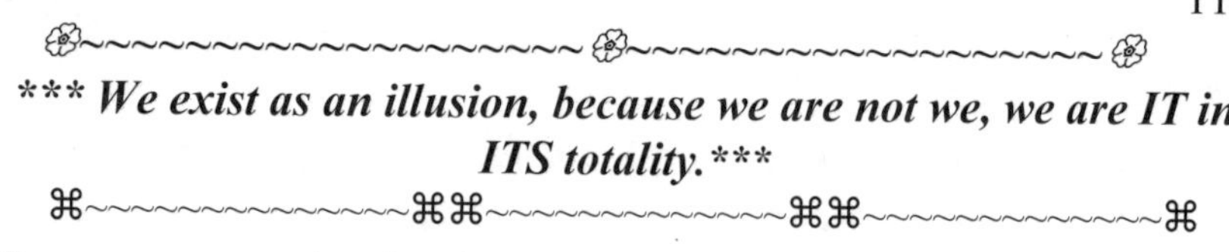

****** We exist as an illusion, because we are not we, we are IT in ITS totality.******

Pure energy and extinction

When ***IT*** completed ***ITS*** mission in the form of dinosaurs, ***IT*** reshaped into a meteor that changed the face of the Earth forever and allowed for the evolution of humans. Already, our number one priority for existence is survival; imagine our lives if dinosaurs still grazed this planet. Would they be supplementing their diets with our sustaining grains or maybe our watery flesh? Would we have evolved to our current status on Earth? I think it's unlikely we would continue to exist.

What we call extinction is actually ***IT*** reshaping ***ITSELF*** in this moment. Our existence happens as ***IT*** reshapes from one Earth rotation to the next. We try to prevent the unnecessary extinction of animals, yet we do not get upset when jobs or material objects become extinct. Well, we may get upset, but it seems rather pointless, doesn't it?

Reincarnation is the same as IT reshaping into something else.

Everything that we understand as having been created eventually undergoes extinction so ***ITS*** reshaping can continue, which is to say that everything created will also be destroyed. However, when all is analyzed, just as nothing is actually created, nothing again is ever destroyed. It is accepted in the scientific community that not a single atom can be added to or removed from the total amount of pure energy in this Universe.

This reinforces the idea that pure energy cannot be created or destroyed; rather, it undergoes perpetual transmutation. But have scientists, have we, honestly viewed this indestructible pure energy as being everything there is? Are we willing to accept pure energy as the One that some have called GOD, others The Creator, and right here, I'm calling ***IT***? For if pure energy is the substance of ***ITS*** existence and ***IT*** can neither be created nor destroyed. Neither can we…

*** *We should be grateful for extinction!* ***

⌘~~~~~~~~~~~~~~~~⌘~~~~~~~~~~~~~~~~⌘

The creation of nothing

In the process of understanding ***IT***, we let go of old ideas about creation and the view commonly held that God created everything. A scientist's definition of creation in fact would share my perspective that the Universe is the result of the pure energy that has always existed; what we see as creation is just the reshaping of energy through transmutation. But even science focuses on the presence of pure energy as the matter that we visually see, and consequently, we have tons of information on the way pure energy exists as matter, from atoms to all celestial bodies. While science used to see matter as 10% of the Universe, now science tells us that the matter of the Universe is more like 4.6% of its totality. And as our minds become more conscious due to our accumulation of information on the existence of God and pure energy, it will be easier to see that regarding God, our existence, and that of this Universe and of life, the word creation is missing something. When in the beginning our minds had first contact with God's Earth, seeing ***ITS*** "creations" as everything that exists as matter, we were not aware that ***IT*** also existed as a cold, clear nothingness. Now add to this what science has found or knows exists as pure energy, for it is the same as God, as omnipresence: Pure energy exists only as this existing moment.

❀~~~~~~~~~~~~~❀~~~~~~~~~~~❀

*** ***IT** **has no beginning or end*** ***

⌘~~~~~~~~~~~~~~~~~⌘

I've already explained why creation is a problematic word regarding the idea that something comes from nothing. Here it remains a problem when we consider that ***IT*** also exists as a constant nothingness or as a constant moment in an area that is not visible as what we see reshaping or that which we generally call Creation. This further shows that what we have been calling ***ITS*** Creation is only that which we can visually see: ***ITS*** weight, as the matter that exists inside of ***ITS*** cold, clear nothingness.

When I have tried to use the word creation in my everyday life, I have found that because I am a result of ***ITS*** ever changing weight, I am unable to find one fragment of me that I can say has been created. Every atom that I exist as is ***IT***, as God, or pure energy, as ***IT*** reshaped or transformed ***ITSELF***.

I tried using the word creation in reference to my children, and here again, the word does not apply because I, in no way, could create the bodies that my children now have; I did not create them. They resulted from ***ITS*** reshaping ***ITS*** weight as ***ITSELF***; this better describes how my children came to exist.

The same applies to all that exists outside of me: homes, cars, and everything else that we refer to as man-made creations. I have noticed that for these so-called human creations to exist, we first had to exist as us, and as us, we have to use our minds to reshape ***ITS*** weight, as matter, so that we can call something like a car, or house, a creation. In order for us to participate in this so-called creation we have to use our minds, and our minds are totally ***IT*** as ***ITS*** weight, as all the atoms that make our minds and bodies possible. Therefore, in order for us to take part in what we are going to call a creation, using our minds and bodies, we have to be fully energized by ***IT*** as ***ITS*** oxygen that ***IT*** exists as, and as the food that we need as fuel in order to participate in what we call creating. I do not see where we are creating anything. This does not bother me, however, for how can I be disturbed by that which exists as everything?

Let me also add that as much as our minds do not have a problem saying that God created everything, they will very seldom include ***IT*** in what we refer to as us creating something. We can see that our minds want to be independent when we refer to what we create as we use ***ITS*** weight as the materials with which to make something, anything.

Here is one more thing for you, the reader, to think about: No one ever created ***IT***, ***IT*** is not a creation.

So, you can see why we have to review this word "create" in order to include ***ITS*** nothingness that exists in everything that is stated as being created by someone (us) who is also a manifestation of something (***IT***) that exists as not being a creation in itself, as in

GOD, or pure energy, not being created.

I can see the human centered understanding of creation as the result of how we have been educated. I learned that it was an admirable thing that we humans create things during our existence. Additionally, it seemed natural that humanity evolved in stages directed by our God, which does lend to ***IT*** some of the respect ***IT*** deserves. But I have also found that using the word create regarding existence and evolution has blocked me from clearly seeing and understanding ***IT*** as the pure, ever present energy ***IT*** is.

The drive that emerges from our feelings about creating likely helps to ensure that the powerful reshaping force continues. In this way, we are instruments of ***ITS*** reshaping. But it is important to let go of this feeling of creative ownership in exchange for a better understanding of ***IT***, and if we are to think of ***IT*** as God, we must add to the definition the existence of ***IT*** as pure energy and as a place of omnipresence.

Only recently have we begun to attach the quality of omnipresence to our notion of God. During the earlier stages of human development, we saw animals as animals and people as people—our minds labeled everything in terms of individuals: an atom is an atom; chemicals are different types of chemicals; planets and galaxies become one more planet or galaxy. But if we accept the oneness of ***IT*** as omnipresent, we will see that nothing is independent of ***IT***; everything exists within ***IT*** as a place of omnipresence.

IT is 100% of all that exists including the trillions of individually named life forms that have inhabited this planet throughout history, and history is only a concept put together by the human brain—a brilliant organ but one made mostly of 85% water. ***IT*** was reshaping long before the brain and the mind came into being, and well before the mind contrived the convenience of the mechanical time system. Because of time, for example, we understand that before our ancestors came into being, certain animals existed and became extinct. The dinosaurs that ruled the Earth several million years ago exist no more. Yet they remain as omnipresent as we are here and now, for dinosaurs were part of ***ITS*** reshaping as this same omnipresent moment.

We exist in ITS image

As a human body, we each are one unit that is 100%. You, for example, as one human being, are one whole unit of human life. Yet you and I, and our civilization are less than a grain of sand in ***ITS*** vast Universe. Consider the fact that millions, billions, and trillions of living organisms exist within our body. These bacteria live within your 100% much the same way the six million humans and the billions, trillions, and googols of other life forms exist within ***IT***. However, this isn't to say that ***IT*** is a giant body or that ***ITS*** origins are a human body. Neither are you a giant bacterium. To exist in ***ITS*** own image means being one within ***ITS*** constant transference of pure energy.

And since pure energy cannot be created or destroyed and life is the result of this pure energy, this should be an indication that life cannot be created or destroyed

❁~~~~~~~~~~~~~❁ ❁~~~~~~~~~~~~❁

** ***All living beings are IT as Pure Energy*****

⌘~~~~~~~~~~~~~~~~~~~~~~~~~~⌘

The devil

If the Devil were to exist, it would have to exist within this Universe. If a devil existed within this Universe, it would have to be part of God as omnipresent. But why is the Devil typically depicted as male? Is it because a male invented it? Maybe it's because women are angels!

Why hell does not exist

There is no Hell because for a physical hell to exist it would have to exist somewhere, and would also have to be made of something; otherwise, we would not be able to recognize this hell as a place. Now, if this hell was made of something then it would have to have some of ***ITS*** weight as the matter that makes this physical hell possible. And if this hell did have a place of existence it would also have to exist within ***ITS*** body, for as the phrase goes: ***IT*** created everything and everything that is exists within this place called omnipresent. At least, this is the way I see ***IT.***

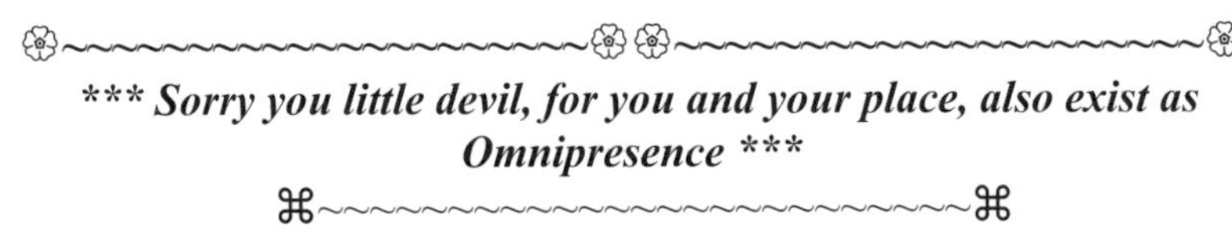

****** Sorry you little devil, for you and your place, also exist as Omnipresence ******

Humans as killers

Most human beings would like to believe that they are not killers. Humans are killers. In primitive times, humans have had to "kill or be killed." Modern humans also kill. Everyone has killed a mosquito, a fly, or that cockroach that was once a living thing. The meat that most of us eat comes from what was a living creature. Someone killed to help provide for your survival. This is the duality of existence: life and death.

During our work, or what some might call the day shift, we could at any moment get into an accident where someone else as ***IT*** will determine if we are to depart as life. Or death might result from our own negligence, maybe in the form of a heart attack. It can also happen without human intervention, as when we get physically tired and are forced to rest, and there in our sleep, ***IT*** does not let us wake or return to work again. In this way, ***IT*** lets us take part in transferring ***ITSELF***. We continue evolving because ***IT*** continues reshaping.

Those of you who have come close to death will agree with me on the following statement: property and material objects have little importance. When we depart this Earth, we cannot take material possessions with us. We should give more of our energy to the people around us. We should enjoy the moment that we have while we are alive.

Have you ever noticed that when you are at a funeral, everyone has something to say about the deceased? The deceased person cannot hear what is being said. I believe that we should say what we want to say to the people that we care about while they are still with us. Death is only a moment away for all of us. We should be grateful for the moment that we have been permitted to exist on this planet as life.

When I die, please do not come and see me. I cannot attend to your needs. I would like for my body parts that are functional to be used by others that may need them. I would also like for the remains of my body to be cremated as I will have no further use for them.

IT is not infinite

In seeing God, the Universe, and everything that exists as one, you will see pure energy at work, as ***IT*** continues to reshape into infinity. ***IT*** is everything that exists, and everything that exists is ***IT***, yet ***IT*** is not infinite. ***IT*** is one, a whole that totals 100%. Infinite are the possibilities of ***ITS*** reshaping ***ITS*** weight. In ***ITS*** majesty, ***IT*** reshaped from very dense matter into this Universe and everything in it—into atoms that reshaped into water that reshaped into human cells that reshaped into organs as ***ITS*** weight.

IT gave us the opportunity to exist as humans here in this place known as the 21st Century, which gives me a very warm feeling. ***IT*** reshaped into our minds capable of enjoying the moment and thinking freely. Everything in the Universe is ***IT*** as a total, perpetually reshaping ***ITS*** weight only, from one form to another while seeking all existing possibilities.

The Universe and God

Imagine the universe when ***IT*** was 99.99% cold nothingness with a very hot and dense center. This dense matter was ***IT*** as part of the total 100%. ***IT*** became opposing forces within ***ITSELF***, both a positive and negative energy at the same moment in the same place. When these energies made contact, they caused the violent explosion that we call the Big Bang. This enabled ***IT*** to reshape into the Milky Way and into the arrangements necessary to form atoms and what we understand as life in this moment.

IT rearranged ***ITSELF*** into a planet that would be just the right distance from a sun that water could be maintained with the proper amount of evaporation, the process we know today as the hydraulic cycle. Water was the onset of our existence. By reshaping atoms into water, ***IT*** made the first living cell possible. And by reshaping atoms into DNA, ***IT*** gave cells the information necessary to develop into every particular creature. So from one cell, ***IT*** reshaped into diverse life forms: first sea bacteria and algae, then species that could see, and hear, move, and taste, species with complex digestive systems and reproductive organs, and ***IT*** continued to reshape into all existing possibilities. This constant change into different types of life is evolution.

By the time God arrived to the point of reshaping ***ITSELF*** as this same moment, ***IT*** has been through several types of eyes, ears, teeth, mouths, and bodies. ***IT*** has now reshaped into nano-technology, so that ***IT*** can do more things with the human body. However, God could not visually see ***ITSELF*** in the beginning of ***ITS*** reshaping. By putting together a human body, ***IT*** was able to reshape using the qualities afforded to human beings: seeing, hearing, and thinking as well as being different sizes, colors, and shapes. By becoming human, ***IT*** could then build the objects that are found only here on Earth as man-made.

IT arranged political systems that span from one extreme to another, and ***IT*** used the same extremes when reshaping into objects such as automobiles, TVs, and VCRs. Look at the size of a spacecraft compared to a house. Look at a microchip compared to a skyscraper. Whatever ***IT*** decides to reshape into is possible. The human form first appeared to resemble the ape. As the human form evolved, so did human life. The temperature ranges on Earth provided humans the chance to wear clothing. Clothing began as something required for functioning: warmth, protection. Clothing has progressed to what we now know as fashion. Different types of fashion are appropriate for different types of roles in this show called life.

PART # 3

Introduction

In this section we will be talking about matter and speed. More than that, we will be talking about all the emptiness that exists as the Universe and inside every atom. This section will help you to accept that everything that exists as pure energy, as God, and as ***IT,*** is and is in omnipresence.

For some this will be easy reading, and for others it won't be. But this is an all-in-one book put together for those who do not understand certain subjects. I hope you re-read the parts that make no sense to you. I hope all readers read this slowly and more than once so that what was already clear may act as a light to see the more shadowy parts. You will likely want to rethink some things that are ordinarily thought of as normal. Try to forget the existing labels that we have placed on this Universe and the way we believe that it should be.

Please take this information as something that will help you understand ***IT*** - pure energy- the creator-God-that which makes everything possible. Do not even look at this text as mere information but rather as an example of ***IT***-pure energy-, reshaped as matter and as weight as ***IT*** plays with speed controlled by electrons. For in this section, we will see, using simple mathematics, that when we take away ***ITS*** weight, this pure energy, as a duality, will add up to 99.99% nothingness as ***ITSELF***, as pure energy.

We can now, with advanced information and technology, understand this nothingness, as a part of that which is called Dark Matter, even though it is not really dark; it is clear, by which I mean transparent. You will read how this emptiness exists in everything, including our bodies. You will also read why light is not an entity itself but a by-product, and here too is why time does not stop at the speed of light, but rather where ***IT*** can move from the Sun to other parts of ***ITSELF*** behaving as a wave and a particle simultaneously.

First however, we shall look at the Big Bang as pure energy: ***ITS*** weight and density, and ***ITS*** spin and speed of 186,000 miles per second (186k mps). We will look at water and our Universe and the three scenarios of ***ITS*** existence including ***ITS*** constant presence, ***ITS*** omnipresence, as nothingness. This is not a Science 101 class, but rather information on ***IT***, which we have been able to gather with the aid of modern science and technology.

Let me also mention to you the reader that we in general see things as having a mass, and this has to be so, because we are visually seeing something as having a mass. For naturally in order to see anything we do need to see ***IT*** as the mass that ***IT*** exist as.

Now let me explain something that will help you understand some of the subjects that you will read about in this section, and that you will better understand what I am writing about, if you remember that for me to understand ***IT*** better, I have had to see ***IT*** as ***IT*** is now, and how ***IT*** existed at the moment of the Big Bang, and even before. but to do this I have had to reduce the many ways that ***IT*** exist as now, that would still be the same, now, and the same at the moment of the Big Bang, and how ***IT*** could have existed before the Big Bang, so please remember that it is only a way of seeing and understanding the way ***IT*** exist, that I present to you what you will be reading, as I see ***IT*** as ***ITS*** duality, 1- is ***IT***S nothingness, and the other is ***ITS*** weight, and ***IT*** is from ***ITS*** weight, that anything that can exist, exist as, and **anything that exist has to exist within *ITSELF***, let me also tell the reader as I have mention before that I am not a chemist, or a physicist, and as what my education is compose of I will elaborate more on later in this book.

But returning to my main point, which is, that everything that I have written in this section, and what follows, is based on a question that I made to ***IT***, which was, that I did ask ***IT*** if ***IT*** would let me understand ***IT*** better.

Now this does not mean that everything that I have mention is an absolute truth, for I too have my limitations, to which Is why I welcome all points of view that relate to ***IT***, for I too know that there is a lot more to learn about ***IT***, that other readers have, as information, and that to me, ***IT*** is not important how ***IT*** exist outside of me, as much as how ***IT*** exist with in me. For I know that ***IT*** is huge, and that we are just beginning to understand ***IT*** better, and that to me, no matter how ***IT*** exist outside of me, in ***ITS*** peaceful ways, and as ***ITS*** violet ways, simultaneously, I will always have to say to ***IT***, thank you for allowing me to be here in observing the way that ***IT*** exist as, and since I do know that ***IT*** did have weight, as in the weight that ***IT*** had at the moment of the Big Bang, and this weight still exist, as the weight that all the matter that now exist with in this Universe ,is the same weight that ***IT***S weight existed before the Big Bang, and that this weight has always existed inside of ***ITSELF***, which is made of a cold clear invisible nothingness, as ***ITS*** body, So that when I say that light has to have some weight attached, is because this light is something, because as I have mentioned before, that since ***IT*** exist as two extreme dualities, One as ***ITS*** weight, and two as ***ITS*** nothingness, so that for anything to exist, as in light, this light has to have a portion of ***ITS*** weight attached, here is one more way to see, what I am talking about, lets go back to the moment of the Big Bang, so I can show you something, at this stage of ***ITS*** existence ***IT*** had all ***ITS*** weight in one place, as weight, now if this high speed of 186,000 miles per second (hereinafter mps), does exist, as pure energy, then this speed has to be there too, because just like ***ITS*** weight, which cannot be created or destroyed, also applies to this high speed (that ***IT*** exist as) that we call 186,000 mps, which cannot be created or destroyed either, and maybe even faster than 186,000 mps. Which is what I gave the name ***MAXX-SPEED*** as existing with in ***ITS*** nothingness.

Here is one more thing that you, the reader, should consider: There is a law in physics that states that if something cannot be disproved then it is possible. But I think that what is more important is that what I have offered you in this section will help you understand ***IT*** better, not as physics or as chemistry, but how ***IT*** exist as omnipresent.

The Big Bang as omnipresent

To understand the Big Bang theory we must first discuss what existed before the Big Bang, before the Universe as we know it began existing. If we were to calculate the mass of all that currently exists in the Universe, the total weight, which is considered to weigh tons per square inch, would have to equal the weight of the Universe as it existed in the form of a very dense pure energy at the moment before the Big Bang.

One truth that I have determined during my research is that the only thing that is one-of-a-kind is this pure energy that existed as dense matter before the Big Bang. This pure energy can be defined as omnipresent. We should be conscious of the term omnipresent and not take it for granted. It is not used frequently in our everyday lives, yet it signifies everything that exists in our lives.

Omnipresence encompasses the secrets of existence before and after the Big Bang. Within the context of the term omnipresent, we come to understand pure energy as ***IT***, as God, and as the Universe. ***IT*** reshaped from dense matter into the Universe, maintaining omnipresence and the conservation of pure energy: nothing created, nothing destroyed.

From the Big Bang and ***ITS*** omnipresence, we know the Universe functions as a duality, an example being the basic hydrogen atom composed of one positive charge (the proton) and one negative charge, (the electron). This duality is made apparent at the very beginning, in ***ITS*** instant reshaping, when pure energy transmuted immediately from dense matter to an energy having dual forces within ***ITSELF,*** both a positive and negative energy. When the positive made contact with the negative, ***IT*** produced a violent explosion known as the Big Bang.

Throughout the Universe, we see this duality in action. The temperature of the Universe provides an example. Temperatures vary from extreme cold to extreme heat. Extremely cold temperatures are found in dark empty space; here ***IT*** has the capability of absorbing energy. Extreme heat is found in light such as that of the Sun where ***IT*** emits energy. Our Sun is said to produce 4% light and 96% heat. Our Sun and all the other stars transfer energy to the Dark Matter that makes up space.

Light and dark, and hot and cold work together. Neither the negative nor the positive energy alone is significant. Their significance exists in the mutual duality of their forces. In this duality there is absolute perfection.

Applying the concept of duality further, let us compare ***IT*** to a battery. The battery has energy stored inside it. This energy transforms into positive and negative charges when inserted into a mechanism requiring power. The battery has not changed; it has transformed itself into dual forms of energy. This is omnipresence in a simplified view. Any form of energy, positive or negative, is merely waiting for the opposite form with which to make a connection. When the connection is made, the stored energy transforms. The meeting of the positive and negative forces causes an explosion that changes stored energy into light and heat.

Light travels at 186,000 mps, a speed at which matter cannot exist; therefore, ***IT*** is not matter at this speed, but rather a duality that is both positive and negative in the same way that electrons can behave as both particles and waves.

Light transfers energy from one place to another. This is how ***IT*** reshapes ***ITSELF*** maintaining omnipresence.

⌘~~~~~~~~~~~~~~~~~⌘⌘~~~~~~~~~~~~~~~~~⌘

*** ***Light as a duality is where IT can move ITSELF simultaneously as a wave and particle.*** ***

❀~~~~~~~~~~~~~~~~~~~~~~~~~~~~~~~~~~~❀

The conservation of pure energy

Consider duality with respect to atoms. Hydrogen is generated in abundance; 90 % of the Universe consists of hydrogen atoms. Stars convert hydrogen atoms into light, and heat. Hydrogen's electron has a mass that is 1,836 times smaller than the mass of the proton. The electron is the outer part of the atom that is traveling near the speed of light. It is considered the negative part of the atom; the atom cannot exist without it. It has an equal force of energy to the proton, but the proton, having the greater mass, is the positive part of the atom.

The proton exists only as long as the electron is circling it; a proton without the electron will decay and transmute into a wave or vibration. Therefore, as a duality, the atom stays within the laws of the conservation of pure energy, which maintains the established scientific truth that nothing in the Universe can be created or destroyed.

Now imagine that we are in outer space in an emptiness that is one great atom of hydrogen that we can actually see. I did say *imagine!* Let's remove the electron, which is supposed to protect the proton, leaving all that remains in the empty space of the atom: nothing. Without the electron, which provides the speed that provides movement-the vibration necessary for material existence-the atom goes into decay. As it decays, its weight diminishes beyond being detectable. Even what we understand as mass, such as the mass of the proton, is then only seen as waves and vibrations.

At the beginning of the formation of matter, ***IT*** combined atoms having masses made of vibrations in order for ***IT*** to move energy around. If ***IT***, as mass, were solid, it would be harder to move from point A to point B as ***ITSELF***.

IT reshaped as matter

Two thousand years ago, we did not understand of what matter was made. Today we recognize that matter consists of atoms. Part of understanding matter in terms of atoms is also recognizing that there are 92 naturally occurring elements made of atoms. ***IT*** as pure energy reshaped ***ITSELF*** to form these natural elements, familiarly classified into what is known as the Periodic Table of Elements. When matter is reduced to its lowest components or building blocks, we will find one or more of these 92 different elements.

Understanding atoms also requires an analysis of their constituents and functions. Commonly taught and learned is that all atoms have electrons, protons, and neutrons. This is not completely accurate, for atoms begin with hydrogen, where matter begins. The hydrogen atom has a proton and an electron, but it has no neutron.

No life could exist without the hydrogen atom. Hydrogen is classified as element #1. It provides sustenance for all living organisms as they go through a series of chemical reactions. These chemical reactions have complex structures such as the molecules of DNA and proteins.

In the nucleus of a cell we find DNA molecules, which combine with RNA molecules, the base upon which protein combinations are formed. Because of the combination generated by DNA, an RNA molecule carries a code that in turn gives proteins a precise sequence of amino acids.

The code that insures the passage of cells from generation to generation is possible because DNA molecules reproduce during cell division. Certain combinations in these structures must be capable of breaking and reforming with great ease. The exact energy needed for this breaking and reforming exists only in hydrogen atoms.

⌘~~~~~~~~~~~~~~~⌘⌘~~~~~~~~~~~~~~~⌘

****** The difference between humans and chimpanzees is the difference of a very small percentage of DNA. ******

❁~~~~~~~~~~~~~~~❁❁~~~~~~~~~~~~~~~❁

You will find the hydrogen atom in outer space because it is the lightest element. Heavier atoms, particularly those heavier than oxygen, which consists of eight protons, eight electrons, and the usual number of neutrons, will be found naturally from the ground down.

Atoms function in an orderly fashion. The electron circles the nucleus to ensure that the proton remains inside. This electron also ensures that nothing from the outside of the atom can get in without the proper conditions that govern atoms. If you add one more proton to a hydrogen atom, and you add a neutron to hold together the two protons, and one more electron to the already existing electron, you have a helium atom (element #2). Each time an additional proton is added, a neutron has to be added; plus, an extra electron is also needed to form a new element.

If you take a hydrogen atom which has one proton and add 7 more protons, and more neutrons to hold the protons in the nucleus, and add the necessary electrons on the outside, you will then have an oxygen atom, which is element # 8.

The periodic table lists the elements in this progression. As the number of protons increases, the heavier the atom becomes. While the elements become heavier and denser, however, they never become solid.

In determining the neutrons necessary in an atom, keep in mind that energy of the same type (positive or negative) repels and opposites attract. When a proton is added to an atom, any preexisting proton will push away or repel the added proton. The neutron counteracts this. Thus, a neutron becomes necessary when there is more than one proton inside an atom. The neutron holds the proton forces together, which enables two protons, for example, to coexist within an atom.

Imagine this: You are a neutron holding hands with two protons. You have one proton on your right and one on your left. With every increase in protons, additional neutrons must be added to keep the protons in place. This example applies to every atom in the Universe.

All matter we know of consists of these 92 naturally occurring elements. Each element behaves differently. Let's look at element #80, mercury. Some of us as children have played with mercury in a science class or seen it function as a temperature gauge in a thermometer. Take one proton away from mercury and the element becomes gold, element #79. Gold functions very differently than mercury.

An atom's function is determined by its electrons. Without electrons, atoms cannot exist. Scientists have discovered that atoms have different electron voltages, ranging from 4 to 24 e-volts, depending on the element. An interesting fact that I just happened to notice is that hydrogen and oxygen, the atoms that form water, have the same electrical voltage: 13.6 e-volts.

The illusion of matter

What gives us the illusion of matter and solidity is color, and the reason we bump into matter is because of the force of the electrons that is holding ***ITS*** weight inside the atoms. When I look at a fruit for example, I know that it has one thing in common with everything else in this Universe; it is 95% empty space. Yet each fruit has a different color, so it is to color that I attribute one reason I cannot see through it.

We can determine that we can see color and touch matter due to the part of the atom known as the electron. Electrons orbit every atom near the speed of light and doing so makes the atom's existence possible. While orbiting, the electron cloud holds atoms as elements together. It bonds with the electrons of other atoms enabling different atoms to bond and form the conditions of matter that exhibit different colors and weights and the resulting illusion of solidity.

For example, atoms bond to form resistant materials such as wood and steel, which have density. Another resistant material is concrete, made basically of cement, sand, and rock. Sand and rocks can bond together with what is known as Portland cement, which is basically made of 64% calcium and clay, and by using steel rebars to act as the concrete's skeleton, it will stand without toppling over or breaking off. Cement is brittle and will snap without the steel rebars that help it to resist breakage. Yet despite all these conditions and materials, concrete is still 95% empty space because the 95% empty space within each atom always remains.
Imagine a concrete wall that is one meter thick and beside it a steel wall that is also one meter thick. Looking at both walls, one would believe that both of them are solid. The walls exist because they are made of matter, weight, and speed, but the densities of the walls will differ according to the elements out of which they are constituted. Yet actually, neither is solid. Each wall is 95% empty space, for each is made of atoms, which are 95% empty space.

Why weight and not mass

I have tried to find ways to express what we understand as mass, which is weight and extension in space. Let me give you an example that relates to our understanding of mass that cannot be applied to ***IT***.

If I take a balloon and fill it up with oxygen, as I look at this balloon I can see its mass occupying space. Because the balloon is here with me, inside the Earth's gravitational field I know that the balloon has weight: Not only the weight of the balloon itself but also the weight of the oxygen atoms that are inside the balloon pushing it outwards.

Now if I apply this as an analogy to ***IT*** as mass, it won't work because since ***IT*** exists as this Universe, (as omnipresent) if I try to see ***ITS*** mass occupying space as something that has a beginning or end, I cannot, because this hugeness that ***IT*** exists as is made from a form of nothingness, and if I tried to weigh ***IT*** as ***ITS*** nothingness, I would be unable to do so because this nothingness does not have weight. For us, weight began increasing with the heated weight that ***IT*** fragmented into the first hydrogen atom's sole proton, and the amount of weight that a proton has can decrease into smaller fragments, such as photons, but both of these are composed of ***ITS*** weight that exists within ***ITS*** one total nothingness.

We will be able to understand ***IT*** better if we think about this in percentages. ***IT*** is a whole; therefore, in ***ITS*** entirety, ***IT*** is 100%. However, ***IT*** exists as a duality of weight and nothingness, in which ***ITS*** nothingness makes up 99%, while ***ITS*** weight is only the remaining 1%. Furthermore, ***IT*** holds this 1% weight within ***ITSELF,*** that is, within ***ITS*** 99% nothingness, thereby making up a whole, or a total of 100%. And I should add that the only reason why I say ***ITS*** weight is less than 1% is because I have had to use the scientific information we have that relates to ***ITS*** weight as atoms (matter).

And as to ***ITS*** weight in atoms, my calculations on ***ITS*** very dense weight as it existed before the Big Bang, do not bring me closer to what we have been discussing. The idea proposed by scientists that this very dense weight just before the Big Bang was tons per square inch does not seem possible.

But again, I have to remind you, the reader, that it makes no difference how huge, or heavy ***IT*** may be, for my personal experience is that it is nice that ***IT*** will let me see and understand ***IT*** better through the scientific information that now exists, but I would not trade this for being as close as I can to ***IT*** by connecting with ***IT*** through meditation, for ***IT*** is here where I can enjoy ***ITS*** existence as me, and as ***IT***, and yet, I do continue to talk to ***IT***, for I know that ***IT*** is there as omnipresence. Among some of the things that I say to ***IT***, are: ***YOU*** (***IT***) are incredible, amazing in the way your divineness, as our life, exists within your nothingness, ***YOU*** are incredible as to what you can do with your weight, as the biggest building, from the biggest jet airplanes to the smallest switches and motors that ***YOU*** are reshaping into through nano-technology.

So, ***ITS*** hugeness is not made of ***ITS*** weight, but does have extension, so let's see what a mathematician can come up with concerning this situation.

IT as three scenarios

IT could exist as three forms of nothingness: One, the way it is now, in which ***IT*** has ***ITS*** weight spread out as hydrogen and the heavier elements in apparently infinite quantity throughout the Universe. The form in which we presently exist as we understand it could be the only reality. Two, is the way ***IT*** existed before the Big Bang, having all ***ITS*** weight in one pinpoint in the center of this nothingness. And three, ***IT*** could have all ***ITS*** weight distributed so finely and evenly throughout the Universe that it cannot be detected or perceived.

We know that ***IT*** exists in the form we are presently observing ***IT***. And we also believe that ***IT*** existed in a very concentrated form before the Big Bang.

This emptiness has always existed, for this emptiness is ***IT***, as ***ITS*** body where ***IT*** has ***ITS*** weight within this nothingness. But there could be a different form. Within this same empty space, ***IT*** could have mass that is not dense and a weight that is not quantifiable; it could be just an infinitely undetectable something.

We should remember that as a proton gets smaller, it becomes harder to find and weigh. So that if scenario #3 did exist, you might see, if magnified, the same empty space as one huge space with nothing in it: clear nothingness. If all the matter in the Universe was distributed evenly and infinitely and if it were undetectably small, throughout all of the space that ***IT*** can exist as, we would not be able to see anything, would we?

Let us say that we have 100 pounds of a very fine, clear powder, so fine that it is microscopic. If we distributed this powder evenly throughout a huge area, we would end up with the same area and the 100 pounds of this clear powder. We would not be able to see it and would hardly detect it, but the weight would be there. So, on a grander scale, if ***ITS*** weight were similarly distributed, it would appear as one huge, clear, nothingness with no beginning or end. But this clear nothingness is ***IT***, as something, because ***IT*** does exist.

All of the above may seem strange, but not any stranger than our perception of space, matter, and time as we understand it now, considering we exist in a Universe that is more than 99.99% empty. If it were not for protons and neutrons (weight) and electrons (speed) that allow matter to exist, we could not even exist.

This third scenario of ***IT*** as an evenly distributed weight is a possibility because ***IT*** loves to practice ***ITS*** ability to reshape into all existing possibilities. From this ability to reshape, we can know that ***IT*** existed in a pre-Big Bang form. And ***IT*** continues to search for and change into new forms as weight. In the same way, our human curiosity is the result of ***ITS*** searching and changing nature.

So if it were not for ***ITS*** weight, ***ITS*** speed, and ***ITS*** emptiness, I would not be here to say thank YOU ***(IT)*** for what YOU are. And as much as I have tried to find a gift for YOU, I have accepted that there is nothing to give, for YOU are everything, even the gift itself.

Straight lines

Before the Big Bang there were no straight lines because it was only after ***IT*** reshaped ***ITS*** weight into matter that we could measure from one point to another as ***ITS*** weight, and when ***IT*** had all ***ITS*** weight in one place, there was no way to measure in a straight line from point A to point B.

The Universe is 99.99% nothingness

On the basis of the nothingness that exists, we may have to reconsider the Universe being 95% empty. ***IT*** could be more than 99% empty space and less than 1% weight as matter.

If what we have been saying is correct, that there is 5% matter in the Universe, then we should be able to understand that because matter is made of atoms that are in themselves 95% empty space, if we reduce that 5% matter by the 95% empty space within that matter, we end up with a total of something like 99.99% empty space.

And if we consider this empty space pure energy, what we are seeing as matter is really just pure energy as weight. You might see this better at the atomic level: We see matter because of the emptiness in every atom--the same emptiness and weight that was ***IT*** before the Big Bang, where ***ITS*** speed of 186,000 mps also existed and still exists in the absence of matter (weight). ***IT*** pushed this weight outwards at the moment of the Big Bang and used the speed of 186,000 mps, along with a little weight, to give birth to the first electron.

IT made the electron an opposite energy so that it could act as a repulsive force to ***ITS*** weight as the proton. And this was done with enough force to maintain a distance between the electron and proton. Since this occurs in that part of ***IT*** that is empty, and exists not as 95%, but as a 100%, this weight is really inside the 100% as ***ITSELF***.

By adding the microweight of subatomic particles to the speed of 186,000 mps, ***IT*** makes the electron negatively charged, and in picking up this mini weight, ***IT*** takes from the 186,000 mps the ability to continue in a straight path. So, in adding weight to 186,000 mps, ***IT*** forces this speed to turn in on ***ITSELF*** so that it can become an electron having a magnetic force with a circular motion, which is necessary to hold that part of ***ITSELF***, which is the weight of the proton.

You might understand this better mathematically. If the 5% that exists as matter is 95% empty space, and you remove the 95% emptiness, then the weight as matter is a total of less then 1%. And since this emptiness within the atom is the same emptiness that we have in the whole of creation, we can understand this pure energy as being more than 99% nothingness because we can combine the 95% emptiness that exists as the Universe and the 95% emptiness that exists in every atom.

And there then remains in existence less than one percent weight, as in all the weight of all the existing protons and neutrons combined. So what we call matter is basically a quantifiable weight that exists in ***ITS*** empty space.

We inwardly know that ***IT*** operates in mysterious ways; the same way that ***IT*** is all the nothingness, ***IT*** is also all of the weight in matter. And my thinking is changing in relation to ***IT*** and ***ITS*** weight. If the weight of the hydrogen atom is a unit of weight that is held together by the electron, which is basically speed, then without this speed this weight will divide into even smaller fractions of subatomic particles as weight (which are more nothingness than anything else). How is one to understand their existence as anything other than a nothingness which isn't really nothing? How much of nothing can you be and still exist as something?

Now I personally know that one pound is one pound, and I know that this pound, as we divide it, gets smaller: ounces, grams, etc. And looking at the way ***IT*** operates, it can exist in even smaller fractions to where our minds may not be able to understand this kind of matter that exists in a Universe that is in ***ITSELF*** made of emptiness.

We know that the hydrogen atom is a very small unit of matter. And yet, as weight, it can float from Earth back into that emptiness that exists as outer space.

Science has stated that before the Big Bang all the mass (weight) in the Universe once existed in such a concentrated form that it occupied a very small area. Now that this weight is scattered throughout the Universe, we cannot even begin to imagine the many things that exist due to the reshaping of the original mass.

⌘~~~~~~~~~~~~~~~~~⌘⌘~~~~~~~~~~~~~~~~~⌘

******Music cannot exist without the silence, as the nothingness, that exists in between the notes.******

❁~~~~~~~~~~~~~~~~~~~~~~~~~~~~❁

The beginning of matter

Matter began with the first hydrogen atom. A hydrogen atom is so small that you would have to line up quite an impressive number of them side by side just to achieve the thickness of a human hair! Going even smaller, the 5% of matter that exist inside the universe is made of atoms with their protons, electrons, and neutrons also consists of 95% empty space. Yes, we can see matter, but which is more important: what we see and believe is there or what is not visible yet nevertheless there? For most of ***IT*** seems to NOT be there.

This is a tough nut for the mind to crack. For us to believe in something, it has to appear as something that we can perceive with our senses, but our reality is made more of nothingness. The good news is that this emptiness that exists as nothing is something; it is pure energy.

Obviously there is much to be discovered by science about the nature of pure energy and ***IT***. But to put things in context let me say that this book is not only about matter but is also about pure energy, for matter, like nothingness, is pure energy. We see this in the scientific equation $E= mc^2$, where E=energy, m=mass, and c^2 as a form of a speeded nothingness. With our present information on pure energy we can update $E=mc^2$ to ***IT*** or ***PE*** (Pure Energy) = mc^2. But there are many who do not think in terms of science. They may see this book as matter but not as having anything to do with energy equations or God. They may know what God (***IT***) means to them, yet still see everything as unconnected. $E = mc^2$ might be clearer to these believers, if it went like this: $G=EMC^2$ (where G stands for God and EMC^2 stands for $E= mc^2$). This way we scientifically connect God to the rest of existence, and we recognize the organic connectedness of everything.

Everything is pure energy, and as such, everything is ***IT***, and everything transmutes in the omnipresent; ***IT*** is what makes every thing possible. ***IT*** and this pure energy are one and the same, 100 percent one.

Let us look at ***IT*** in a mathematical equation. The sum of the parts of the Universe is 100%. Five percent consists of matter: planets, stars, everything else, including us. This 5% is made of atoms that are 95% empty space. This means that the 5% we call matter could be reduced to less than 1% as matter.

The remaining 95% of the Universe, which is called Dark Matter, is where 90% of all hydrogen atoms are generated, so if 95% of the Universe is empty space and 90% of hydrogen atoms that occupy this empty space are 95% empty space then this volume of the Universe is also less than 1%. I am sure that as ***IT*** allows the human mind understand more about ***ITSELF***, we will at some moment understand more about this energy that exists as Dark Matter. I believe, in fact, there are readers that have more advanced information on this Dark Matter that exists as empty space, and if you do and you do not mind sharing it with the rest of us, I've provided you contact information at the end of this book.

⌘~~~~~~~~~~~~~~~~⌘⌘~~~~~~~~~~~~~~~~⌘

*** ***Gravity is the process whereby ITS weight, as weight, is looking to reunite with ITS other fragmented weight again.*** ***

⌘~~~~~~~~~~~~~~⌘~~~~~~~~~~~~~⌘~~~~~~~~~~~~⌘

ITS nothingness as weight

When I look at ***ITS*** cold clear nothingness, there is no light, yet maybe there is, like in ***ITS*** heated weight. Allow me to explain my reasoning: When I look at a fire at night I can see light, which may mean that ***ITS*** light maybe hidden within ***ITS*** heated weight.

ITS weight as heat

While 99.99% of the Universe is empty space that as far as we can measure has no weight, the other part of the duality is very hot matter having a concentrated weight. ***IT*** exists as weight in every atom. Matter, in ***ITS*** movement as stars, generates the heat and light we see and feel throughout the Universe

Why the Universe is not infinite as God

To begin, we should remember that ***IT***, as God, as pure energy, is just one, as when we say one of a kind, or as ***IT*** being 100%

Now as I have explained elsewhere, it is ***ITS*** weight that can change into infinite possibilities.

However, this is not so of ***ITS*** nothingness, for this nothingness is ***ITS*** body where ***IT*** keeps ***ITS*** weight with n ***ITSELF.***

Now when we view the Universe we see it as infinite only because what we are seeing as infinite is the way the body of ***ITS*** nothingness exists, as in not having a beginning or end, but this applies only to ***ITS*** nothingness body, because as ***ITS*** weight, this weight is calculable and we know that ***ITS*** weight once existed as a singularity within ***ITS*** nothingness, as a very dense weight, and we have been keeping track of ***ITS*** now fragmented weight, as the weight of any thing that can exist.

So even if we see ***ITS*** cold, clear, transparent nothingness body as appearing to be infinite, it is not, because ***ITS*** body is just one body. However, since this hugeness that ***IT*** exists as is a form of nothingness there is one thing that we can grab onto: As much as ***ITS*** huge body (the Universe) does exist as what we call empty space, it does have one true quality which is ***ITS*** huge body's freezing cold temperature.

So, even if we cannot see or touch this part of the way ***IT*** exists, we know that ***IT*** is there as the freezing cold temperature that now exists as this Universe. Nevertheless, I wish I could know more about how ***IT*** can exist as two individual temperatures simultaneously as one, because both of these temperatures, hot and cold, are ***IT*** and both exist as this pure energy which cannot be created or destroyed.

I must confess that I too was not used to seeing my God as composed of two extreme temperatures, but science does confirm the existence of these two temperatures, as the pure energy that now exists as this Universe.

I know that it is hard for the human mind to grab onto something that exist as a nothingness, but ***IT*** does exist, as this coldness, while at the same time we know that as ***ITS*** weight ***IT*** has heat, and is less than 1%, and this heat can only exist inside of ***ITS*** freezing cold, clear, transparent body (the Universe). It is clear then that ***ITS*** nothingness has energy also as coldness.

Here is one more way to picture ***ITS*** nothingness, which is omnipresent: As you look inside the Universe at the matter that exists inside it, let's say from one planet to another, one thing you will not notice is the nothingness that exists in between these planets and this is so, because our minds cannot see what exists in between these planets because it is transparent. Now, the way we know that this nothingness does indeed exist is because **this nothingness does have something, and that something is temperature.** Furthermore, omnipresent means that ***IT*** is in all places at the same moment, therefore, ***IT*** is also there as this huge place (the Universe) that does exist as a freezing cold temperature.

Another thing that will help us understand ***ITS*** body is this: When we look at a human body we see that this body exists because it has weight and the weight is inside, but scientifically, what we are seeing are fragments of ***ITS*** weight that exist within ***ITS*** nothingness, for we should never forget that 95% of us and everything else that exists or may exist is composed of ***ITS*** nothingness.

To say that ***IT*** is in all places as the same moment means that this place that we call the Universe is there also, and this means that ***IT*** is there also, as the cold, clear temperature that ***ITS*** body exists as, the pure energy that is there.

So if we wish to say that the Universe is infinite, we are referring to the manner ***ITS*** freezing, cold nothingness appears to our eyes. But let us remember that as pure energy, nothing is being created or destroyed, so that as the freezing, cold temperature that this Universe exists as, it too cannot be infinite, for it does not get bigger or smaller. ***IT*** may seem to be expanding but this is an illusion. This is evident when one considers that we know that when ***IT*** fragmented ***ITS*** once singular weight, these fragments became extraordinarily tiny in order to form atoms, but all of these infinitely smaller fragments when added up, must be equal to the same amount that ***ITS*** weight existed as when ***ITS*** weight was one singular weight, as the pure energy ***ITS*** weight exists as.

⌘~~~~~~~~~~~~~~~~⌘ ⌘~~~~~~~~~~~~~~~~⌘

****IT does not get bigger or smaller as ITSELF. ****

❁~~~~~~~~~~~~~~~~❁❁~~~~~~~~~~~~~~~~❁

Now the opposite of ***ITS*** fragmented weight is ***ITS*** cold, transparent, freezing cold nothingness, and this part of ***ITSELF*** is not getting bigger or smaller simply because it consists of a nothingness, because if ***ITS*** cold, freezing nothingness were getting bigger then where is ***IT*** getting this additional pure energy, and if ***IT*** were getting smaller, then where is the missing portion going to?

In conclusion, when we say that ***IT*** is in all places at the same moment, we have to also include this clear, freezing cold temperature as a place, that ***IT*** also has to exist in as omnipresence.

IT floats as weight

IT can hold ***ITS*** weight; ***IT*** floats; everything floats: planets and all, within ***ITS*** total nothingness. Even the weight that is there has no force in and of itself. When something falls, it does not fall right, left, up, or down in the nothingness we know as the Universe. As we map the Universe, we might find that we might be going around in circles. ***IT*** doesn't have walls, streets, or traffic signs to tell us where we are in this nothingness. There are no believable indications of whether or not we are traveling in a straight line, running around in circles, or just jumping up and down. We may never be able to see ***IT*** as having a beginning or end as we journey inside it. But, we must remember that we exist and are moving through ***IT***, and ***IT*** is actually a vacuum that carries ***ITS*** own weight. Now if that does not simplify your life, I don't know what can! As I see ***IT***, there really is no problem in our getting lost in this Universe; we will always be within ***IT***.

Imagine that you are the size of a hydrogen atom and are inside a blood cell, moving around inside of it. You are floating in the emptiness. No matter what infinity of directions you take, you will only find yourself floating, moving within ***ITS*** interior. When you get to the outer edges, you still cannot leave the cell; you just continue to move endlessly. There is so much space in the cell that a whole lifetime would not be enough to see it all, even if you were traveling at the speed of light. An ant would have a better chance of seeing every inch of our planet than you the whole of the cell.

We can see ***IT*** floating here on Earth as the rain floats in our skies. As much as I know that water has weight, absolute tons of it, it moves around freely as rain. What amazes me is that even as weight, ***IT*** floats.

⌘~~~~~~~~~~~~~~~~⌘⌘~~~~~~~~~~~~~~~~⌘

****** We can never get lost, if we remember that we will always exist within IT as ITSELF. ******

❁~~~~~~~~~~~~~~~~~~~~~~~~~~~~~~❁

ITS weight as gravity

Wherever we see ***ITS*** weight as matter, we will notice that whatever the object, it exists as weight from the outside inwards. This is the way ***ITS*** weight exists in the stage we are in now. It is as if ***ITS*** weight is always looking to return to ***ITS*** oneness as weight, as when ***ITS*** weight existed just as one totality, as in the way ***IT*** was before the Big Bang. I will assume that is why the force of gravity exists, as in gravity being the force that is trying to reunite ***ITS*** weight again into one total dense weight.

Furthermore, ***ITS*** nothingness is where ***IT*** keeps ***ITS*** weight. As an example, all celestial bodies float within ***ITS*** nothingness just as the weight that exists in all atoms, and the weight inside the atom also exists within ***ITS*** nothingness. And ***ITS*** nothingness exists as ***ITS*** outward existence while all of ***ITS*** weight exists inwardly, as within ***ITSELF*** as ***ITS*** nothingness.

Take ***ITS*** weight as a proton; there is no nothingness inside of the proton as ***ITS*** weight, or at least not in this stage of ***ITS*** existence. When we look at ***ITS*** weight as the proton, ***ITS*** nothingness is on the outside, not the inside. When we leave ***ITS*** weight as the atom's proton, we are back to ***ITS*** nothingness as an outside, for the electron is more composed of ***ITS*** nothingness in order to act as a barrier separating it from the next atom that may exist nearby, which is also separated by ***ITS*** speeded electron to allow a separation of ***ITS*** same nothingness for more of ***ITS*** fragmented heat to exist as matter.

⌘~~~~~~~~~~~~~~~~⌘⌘~~~~~~~~~~~~~~~~⌘

****** Gravity is the force of ITS weight pulling, trying to unite ITS weight again. ******

❁~~~~~~~~~~~~~~~~❁~~~~~~~~~~~~~~~~❁~~~~~~~~~~~~~~~~❁

An interesting problem and a thank you

This is for the pros who are interested in finding how close we can get to knowing ***ITS*** weight and size.

The problem is this: the whole of ***IT*** is a hundred percent; less than five percent of ***IT*** is matter. Matter has weight because of its density. For example, a hydrogen atom is 1,836 times heavier than an electron, so we know that it is 1,836 times heavier than something. This something, known as the hydrogen proton, is calculated to have an atomic mass of 1.00794. If matter is five percent of the total weight of ***IT***, this means that at least this five percent has weight multiplied by 1,836. This could be a way to calculate how much ***IT***, as pure energy, weighs.

There is a lot of information to work with, available to all. No one really owns it. I just see it as a way of getting closer to understanding more of who ***IT*** is and how ***IT*** functions.

To me, to be able to see and understand something so majestic, so powerful that can range from the violence of the Big Bang to the tenderness evoked by a new born child is a humbling and much more interesting experience than anything I can think of.

So I thank ***IT*** for permitting me to stay close to ***IT.***

Glorifying ITS weight

We glorify what we see as matter. This means that we glorify ***ITS*** weight, or in other words, that part of ***IT*** that is the less than one percent of ***ITSELF***.

⌘~~~~~~~~~~~~~~~~⌘⌘~~~~~~~~~~~~~~~~⌘

******That things are not what they used to be is because IT is reshaping ITS weight ******

❀~~~~~~~~~~~~~~~~❀❀~~~~~~~~~~~~~~~~❀

We have glorified ***ITS*** marvels, as what God created, or as what pure energy is doing as ***IT*** transforms, or transmutes, or reshapes as ***ITS*** weight, yet we seldom glorify ***ITS*** huge 99.99% empty nothingness that ***IT*** mostly exists as. It is in this nothingness that ***IT*** exists as a life force and as a consciousness.

Now that I have accepted that God exists as this place that we thought God created, that is matter but mostly nothingness, I have to refocus my attention from what I, like many, saw as being God, as something visual or sensory rather than one single entity having fragments of weight within a great nothingness. From this moment on, I will also know God to be the cold, transparent, speeded nothingness that I cannot see. I will also stay close to ***IT*** as that which exists within me, that I can connect to during meditation. Luckily, ***IT*** did not send us to the other end of ***ITSELF***, as this Universe, to find or connect to ***IT*** as ***ITSELF***.

Again, I say "thank you" to ***IT*** for keeping us so close even though we have no idea as to how ***IT*** really exists.

⌘~~~~~~~~~~~~~~~~⌘⌘~~~~~~~~~~~~~~~~⌘

****** Change is the only thing that does not change because change is IT as IT evolves.******

~~~~~~~~~~~~~~~~~~~~~~~~~~~~

IT as change

IT exists as one single entity, and as this one, ***IT*** does everything in duality. One part of ***IT*** is a constant emptiness, and the other is weight (matter). Weight is also constant, in as much as there is always the same amount; but it is constantly changing, reshaping. Weight is what is used to generate the constant change in the Universe.

IT as the friendly ghost

We could think of ***IT*** as a ghost because ***IT*** is there, but as a totality, ***IT*** exists as something that we cannot see except as ***ITS*** weight that is moving about as matter, which is what we have come to know as change.

It helps to compare ***IT*** to a ghost because, likewise, ghosts don't visually exist. Of course, at times we are scared of ghosts. And, at times we are scared of the ways ***IT*** produces fast and dramatic changes inside and around us, such as when it reshapes into accidents, disasters, and other dangerous events, or even when ***IT*** feels that we should awaken or change from the slumber that we may be experiencing to an alert mode.
~~~~~~~~~~~~~~~~~~~~~~~~~~~~

But ***IT*** no longer scares me, at least not to death. I have given up on being afraid of other humans, for I now feel that I am in the safest hands that could ever exist, so ***IT*** is now my friendly ghost, moving about and producing changes with ***ITS*** weight.

IT has these features:

* Weight - because matter has weight attached to it.

* Distance - because the Universe is light years across.

* Speed - because ***IT*** travels at the speed of light.

* Sound and color - because ***IT*** produces waves and vibrations.

* Temperature - because of cold nothingness and heated matter.

* Spin - because matter is not possible without the spinning of electrons.

These features help to explain ***ITS*** existence as omnipresent. They also help to explain ***ITS*** constant reshaping of ***ITSELF*** in order to maintain the conservation of pure energy. Since we have established that nothing is solid, it becomes easier to understand how ***IT*** can constantly reshape ***ITSELF***.

⌘~~~~~~~~~~~~~~~~~~⌘⌘~~~~~~~~~~~~~~~~~~⌘

Anything that exists must have ITS weight

❀~~~~~~~~~~~~~~~~~~~~~~~~~~~~~~❀

Speed and weight

When talking about speed and weight in reference to matter, I would like you, the reader, to consider the following: Since ***IT*** is both speed and weight, and therefore matter, it takes very little of ***ITSELF***, as weight, to make the hydrogen atom. ***IT*** then uses ***ITS*** speed of 186,000 mps to become slower as the electron. This weight is so minuscule that ***IT*** can speed around the weight of the proton at a few million meters per second. In doing so, ***IT*** produces what could be described as a cloud, which acts as a buffer, or better known in the scientific community as a magnetic field. And it is this magnetic field which does not allow us to go through the emptiness, or nothingness that ***IT*** exists as, when ***IT*** exists as atoms.

ITS varied oneness

Before I go on, when I refer to ***IT*** as a oneness I am referring to ***IT*** when ***IT*** exists as both ***ITS*** dualities, as ***ITS*** weight and as ***ITS*** speeded nothingness simultaneously. This is where both of the forces of ***IT*** exist as one. Let me give you some examples:

1 - As that which transfers:

Light is where ***IT*** exists as a oneness, as in light being made of a speeded nothingness and at the same moment as particles (photons), as the weight which ***IT*** uses to transfer weight elsewhere within ***ITSELF***, thus transferring ***ITS*** weight within ***ITS*** speeded nothingness.

2 - As a builder

As the electron, which is where again ***IT*** uses both ***ITS*** forces, as the speeded nothingness of the electron, and as the weight that the electron has as a particle. ***IT*** uses this oneness to hold in ***ITS*** weight in the form of protons and neutrons within ***ITS*** nothingness, so as to form atoms (matter).

3 - As ***ITS*** total oneness

This is where all ***ITS*** weight is evenly distributed within ***ITS*** nothingness, which is what could be happening now as this Universe, as the matter that exists, and is moving farther away into ***ITS*** nothingness in smaller and smaller fragments, where these fragments will become so small they incorporate evenly into ***ITS*** total nothingness, which may be the way ***IT*** will return to ***ITS*** stage 3.

Here is an interesting idea to think about: it is when ***IT*** used ***ITS*** dual forces that we came to exist as life, starting from the Big Bang, which is where we also got our Sun, so that there would be light, such as the light that now exists inside ***ITSELF*** as this Universe, which is why we can see ***IT*** as the things that ***IT*** reshaped into as ***IT***S weight as matter.

Think about this also: If there were no light in our existing Universe, we could not confirm that ***IT*** did exist, for we could not see anything as existing. So in a way, we can say that light exists so we can confirm that ***IT*** does exist.

⌘~~~~~~~~~~~~~~~~~~⌘⌘~~~~~~~~~~~~~~~~~~⌘

And then there was light, so that we could see IT as ITS duality

~~~~~~~~~~~~~~~~~~~~~~~~~~~~~~~~~~

ITS weight as distance

Here are some ideas that will help us understand ***IT*** better as far as ***ITS*** weight is concerned. Let me start with ***ITS*** weight inside the atom, which is the proton. Between the proton and the electron there is a huge empty space that could be from 95% to 99% of all the room occupied; this exists as part of ***ITS*** total nothingness. We also know that ***IT*** once had all its weight in one place before the Big Bang occurred, and that it was from there that ***IT*** fragmented ***ITS*** weight as protons, so ***IT*** could exist as atoms.

I have said the above because I try to envision the way ***IT*** behaves so that I can understand better how ***IT*** reshaped ***ITS*** less than 1% weight that existed as one chunk of weight, and then how ***IT*** divided this one chunk of weight into all the weight that now exists as matter.

As I have mentioned elsewhere, if we could find out what ***ITS*** total weight is in terms of the extension ***IT*** occupies, we would then have a better idea of how big ***IT*** may be. Another way we might begin to understand ***ITS*** extension as the nothingness that ***IT*** exists as is by dealing with what we have - the distances that exist within ***ITS*** weight.

If we knew what all ***ITS*** weight occupied as distance we would have a better idea of how big my boss (***IT***) is. I have accepted that ***IT*** is huge, as far as distance is concerned, yet ***IT*** exists within my tiny body also. But let me come back to ***ITS*** weight, in terms of extension, for I would like you to see something from a different angle. Remember that all the emptiness that exists within this Universe is just one, as in the empty nothingness that ***IT*** also exists as. So try looking at things as if all that exists are just fragments of ***ITS*** once total weight.

When ***IT*** fragmented ***ITS*** weight after the Big Bang, ***IT*** divided its weight into individually quantifiable, fragmented, and separated weight that has to exist within ***ITSELF*** as the nothingness that ***IT***
~~~~~~~~~~~~~~~~~~~~~~~~~~~~~~~~~~

exists as. When ***IT*** divided ***ITS*** weight after the Big Bang into individually quantifiable fragments ***IT*** used ***ITSELF*** as one of the ways ***IT*** also exists as, as ***ITS*** oneness, as electrons, so that ***IT*** could maintain ***ITS*** now individually quantifiable fragments of weight separated as atoms.

Look at the above this way: All the emptiness that exists in atoms is the same empty nothingness that ***IT***, as pure energy, exists as, as one single entity because the distance that exists from atom to atom as empty space is all just one huge empty space that exists as ***ITS*** nothingness.

What we have assumed the different spaces that exist in each atom as being independent spaces within each atom, but they are really one, same, empty nothingness that ***IT*** exists as, where it placed its weight and separated this weight with its oneness in the form of electrons.

Perhaps you may visualize it better this way: All ***IT*** has done is take ***ITS*** individually quantifiable fragmented weight and separated these fragments with a distance within ***ITS*** total empty nothingness.

If you look closely, using your imagination, this is what is happening: A hydrogen atom has one proton and one electron. The proton is where ***IT*** separated ***ITS*** individually quantifiable fragmented weight with distance within ***ITS*** one total nothingness, which we refer to as the empty space that exists inside atoms, and ***IT*** keeps this separation with the force that ***IT*** exists as the electron. This is the way ***IT*** forms atoms.

Now, when you look at objects, remember that, as things, they are made of atoms, as matter, as ***ITS*** fragmented weight that has a certain distance within ***ITS*** total nothingness, as the space that exists in every atom.

⌘~~~~~~~~~~~~~~~~⌘⌘~~~~~~~~~~~~~~~~⌘

****The heat we feel from light is a fragment of ITS clear heated weight***

❀~~~~~~~~~~~~~~~~~~~~~~~~~~~~~~~~❀

Now imagine that you are looking at a supposedly solid square 100 foot iron cube. Remember that it is made of matter, the smallest components of which are ***IT*** existing as atoms. The importance here is that all atoms exist as having a nucleus where the protons (and neutrons, if any) are, as ***ITS*** fragmented weight, and from this fragmented weight ***IT*** established a certain distance within ***ITS*** empty nothingness, which is what we call the empty space that exists inside every atom.

Knowing the above scientific information concerning the composition of matter as atoms, if we now look at the 100 foot dense iron cube, which we consider to be solid, we will know that it is not solid at all, because we can now understand the way ***IT*** reshaped as the things that exist as matter within ***ITS*** weight, within ***ITS*** nothingness, as the empty space that exists within matter.

If you stand in front of this huge, 100 foot iron cube which is supposedly solid, and you start by looking at the left corner of the bottom and look across the bottom to the right end until you get to the right side, you will find that it is really an illusion of solidness. I think I can explain it better, but you will have to use your imagination along with the scientific information concerning the composition of matter: As you see this illusionary iron building from the bottom up, what you are scientifically seeing is a structure that is made up of atoms. As you look across this structure you have to remember what it is made from matter that consists of iron atoms. From left to right, this building exist as ***ITS*** tiny fragmented weight, separated within ***ITS*** huge distance, for this structure has weight attached to it in order to exist, but this weight exists as independent fragmented weight, as atoms, which are in themselves separated. I do mean a huge distance, from one fragmented weight to another because scientifically we know that ***ITS*** weight, as the proton or protons in any individual atom, has this huge emptiness that exists inside atoms.

So that if we now look at this supposedly solid iron structure, which really exists as ***IT*** as ***ITS*** weight, we can recognize the ways that ***IT*** reshapes ***ITS*** weight into this iron cube. ***IT*** took its fragmented weight, as atoms, to make up the material substance of the iron, but here we should remember that this weight is separated, not so much as there being a distance between each individual atom and the next, for we should remember that the nothingness that ***IT*** exists as is only one total distance, not individual pieces of distance as what we think exists as the individual emptiness (nothingness) that all atoms have, or exist in, as in all protons and neutrons having to exist within ***ITS*** one, total, same, empty nothingness.

For this one total distance does exist as one entity, as the total pure energy that now exists. So that as we imagine this structure, try looking at it differently: In the case of this 100 foot iron cube ***IT*** took individual fragments of weight from ***ITSELF*** and arranged them up, down and across, as atoms.

Please keep using your imagination, for as we look at the 100 foot iron cube, you have to remember that this cube is made from atoms that we cannot visually see. But they exist as pure energy, as ***IT***, as ***ITS*** weight, as atoms, made up of protons, electrons, and neutrons and the protons are separated from the electrons and from the neighboring atoms by a huge distance. When I say a huge distance, I am speaking relative to the size of these particles.

To continue, as we now look at the 100 foot iron cube, we can imagine it as what ***IT*** exists as scientifically, as ***ITS*** weight, and through this we will see that ***IT*** placed ***ITS*** weight, as atoms, which are really just mostly empty space, in rows, let's say, from left to right, so that what is really happening is that ***IT*** placed a very tiny amount of ***ITS*** weight, starting from the left side as atoms of iron, one after the other side by side, and ***IT*** is holding this weight, as atoms, together by the way ***IT*** also exist as ***ITS*** oneness, as in the electron, so that ***IT*** could hold ***ITS*** weight as individual fragments within ***ITS*** huge empty nothingness, close enough for ***ITS*** weight to exist, but separated enough, as the huge distance that exists as ***ITS*** weight, as the protons, within ***ITS*** empty nothingness, which is maintained within this empty nothingness by ***ITS*** way of existing as ***ITS*** oneness, as the electron.

And ***IT*** also uses ***ITS*** oneness as the electron, as an incredible force that can hold together ***ITS*** fragmented, distanced weight, for it is this force that ***IT*** exists as ***ITSELF*** which is what ***IT*** is using to hold together ***ITS*** distanced, fragmented weight that we call atoms (matter).

Now try and see if you can imagine this 100 foot iron cube scientifically as follows: ***IT*** has, as ***ITSELF***, as the electron, the force that ***IT*** uses to hold together every piece of ***ITS*** once one total weight that is now separated into distanced fragments that are also separated from one another within ***ITS*** existing total nothingness, and holds these distanced fragments of ***ITS*** weight, by ***ITS*** way of existing as ***ITS*** oneness (the electron). I say one, for it is in this particular oneness that ***IT*** uses to hold together ***ITS*** properly fragmented weight (atoms) so that ***IT*** can exist as you me and that huge 100 foot tall iron cube structure. If we could see this huge iron structure for what it really is composed of, we would be able to understand that this whole structure is ***IT***, as pure energy, as every atom that this cube exist as. What is holding this building up and across is ***ITS*** oneness as the electrons that are holding each individually separated fragment of weight, as the huge distance that exists, as the 95% empty nothingness that exists in every atom.

IT is able to hold these distanced, separated fragments of weight (the protons and neutrons) together using ***ITS*** oneness (the electron) so that this 100 foot cube can exist as we see it, as ***IT*** has reshaped ***ITS*** weight within ***ITS*** nothingness. But you should also see *yourself* as this way that ***IT*** uses to reshape into, as you, and every thing that now exists as matter.

The proper way to speak about this 100 foot solid structure is to refer to it as being not solid, but dense, that is, as ***IT*** having more of ***ITS*** fragmented weight in one place. An example would be when ***IT*** placed less of ***ITS*** weight in one place and separated this minute weight within ***ITS*** total nothingness, and held this minute weight in place by ***ITSELF*** as the oneness that ***IT*** exists as in the electron. We can begin to understand ***ITS*** fragmented weight by studying the Hydrogen atom (#1). And as ***IT*** kept placing more of ***ITS*** weight in one place, then came the Helium atom (#2), then Lithium (#3), and as ***IT*** keeps adding more of its fragmented weight we find Beryllium (#4), and more weight as Boron (#5), and as Carbon (#6), which is where ***IT*** placed more of ***ITS*** fragmented weight to make Nitrogen (#7), and then the most important weight that ***IT*** exists as for our survival, Oxygen (#8), which we take in for our breathing and release as Carbon (#6), joined to Oxygen (#8), in the form of Carbon Dioxide.

But let me continue with ***ITS*** fragmented weight to where the weight is heavier, as the element #26, which is Iron, or when we wear ***ITS*** heavier weight as Gold (#79), or when we use ***ITS*** fragmented weight as Tin (#50) to store food for our survival, or as the way ***IT*** weighs as the Potassium (#19) in bananas. And these fragmented weights can go higher than #100 in man-made elements that don't exist in nature but have been manufactured in the laboratory.

The thing to always remember is that it is ***ITS*** weight that is getting heavier, which we refer to as being denser, as ***IT*** having more fragmented weight in one place within its nothingness.

The iron 100 foot cube is not solid; it is denser, as ***ITS*** weight that exists within ***ITS*** empty nothingness. Now you will also understand if I told you that the 100 foot iron cube is somewhere in the area of being 95% hollow, that is, mostly empty nothingness because of the 95% emptiness that exists inside atoms. The 100 foot iron cube, in terms of its height, is held together with ***ITS*** way of existence, as its oneness, as the electron. This iron cube is ***ITS*** fragmented weight from corner to corner as individual fragments of ***ITS*** weight that are separated, as the huge distance that exists as the emptiness inside atoms, which can produce the most miraculous illusion of something like the 100 foot cube which looks solid but is really just distanced fragments of ***ITS*** total weight within its total nothingness.

So that what we refer to as being heavier, or denser, is because this is where ***IT*** has placed more of its weight as heavier fragmented weight in one place as the heavier atoms that compose that iron structure.

You too, as 100 lbs of body weight, are composed in this same way that ***IT*** uses its fragmented, distanced weight as atoms which are not as dense as the iron structure. You exist as matter, so the next time you look at someone in front of you, think of them as existing as ***ITS*** distanced separated weight, and as you look at their face remember that the weight that face exists as is because ***IT*** can fragment its distanced weight all the way to the top of their face. Scientifically, what you and the iron cube exist as are trillions of individual, distanced fragments of weight, as protons and neutrons with a huge distance from one to another held together by a way that ***IT*** also exists as; a oneness as ***ITSELF***, as the pure energy that ***IT*** exists as, in the electron.

I have no doubt that ***IT*** that can move its one total weight into individual, distanced, fragmented weight as atoms to allow for these temporary marvels that exist as ***ITS*** weight, that now exist within ***ITS*** nothingness as you and me and this Universe.

⌘~~~~~~~~~~~~~~~~~~~~~~~~~~~~~~~~~~~~~⌘

Everything that exists has ITS weight and ITS divine nothingness

❁~~~~~~~~~~~~~~~~~❁❁~~~~~~~~~~~~~~~~~❁

Why does IT arrange and disarrange ITS weight again?

In thinking about this question, I turned to the way we exist. Since the beginning of humanity, as ***IT*** first reshaped into us, we did not have much information as far as experience goes. But now, ***IT***, as us, has had the experience of reshaping ***ITSELF*** from the horse and buggy all the way to the complex exterior industrialization that ***IT*** has permitted us to take part in. And if you are wondering why I said "exterior" industrialization, that is because ***IT*** has already begun the industrialization of our interior with motors and on and off switches, just to name a few of the things being accomplished in the world of nano-technology, where another type of rearranging of ***ITSELF*** will take place. Maybe ***IT*** will allow me to try to write another book on this subject, with the title *The Nano-World That **IT** Exists As*, for this nano-world technology that ***IT*** has started is just at ground level, so that if something like ***IT*** can take us from being cave-dwellers all the way to this high-tech society into which we are just beginning to evolve, then imagine what ***IT*** could do with us as ***ITSELF***, in the nano-world. However, we won't necessarily need to imagine ***IT***, for ***IT*** has already started ***ITS*** construction of this world within our human bodies.

But let me get back to ***ITS*** constant rearranging of ***ITSELF***, using what we call thinking, since we have learned and changed from what and how we were in caves to now being in outer space, as ***ITS*** weight.

I am going to assume that ***IT*** too is constantly learning more as to how to reshape ***ITS*** heated weight, for we, like ***IT***, cannot just exist; we have to continue changing as our weight, as ***ITS*** weight.

One reason why I say that everything that we are going through is new to us as ***IT,*** is because ***IT*** can only use ***ITS*** weight to reshape ***ITSELF*** as just one existing moment, known as omnipresent; so that every time ***IT*** uses ***ITS*** weight to reshape into something else, ***IT*** has to use ***ITS*** one total weight again and again, for ***IT*** is not getting any fatter or any skinnier in terms of weight.

So I asked myself: Did ***IT*** ever know what ***IT*** could become when ***IT*** began reshaping? I thought of the whole process, from ***IT*** reshaping ***ITS*** weight, starting as one singular weight, and then going on to fragmenting this weight to become atoms and then to become matter, and as matter to become a planet like Earth, where ***ITS*** weight could exist, for example, as water, so ***IT*** could exist as the oceans, where ***IT*** could continue to reshape ***ITS*** once one singular weight into life, becoming the many life forms that exist in the sea, as ***ITS*** own image of ***ITSELF,*** as ***ITS*** **one life:** as ***ITS*** weight, and ***ITS*** nothingness.. All this ***IT*** accomplished so that ***IT*** could have mobility, and use this mobility to continue ***ITS*** reshaping of ***ITS*** weight, and be able to leave this ocean, and become the different possibilities that ***IT*** could exist as ***ITS*** weight and ***ITS*** nothingness, and as life, to become the birds of the air. After this ***IT*** continued, reshaping ***ITS*** weight so that life could leave the ocean and come up onto the land. (For we should not forget that we as life exist because of ***ITS*** weight as atoms that also have ***ITS*** nothingness inside, and that we exist as life, because ***IT*** exists as life) Finally, once on land ***IT*** reshaped ***ITS*** weight into the way we existed when we were dwelling in caves all the way up until how we now exist.

It is fascinating to contemplate how we exist now, having the ability to see, and touch, and read about the many ways ***ITS*** weight exists on this planet, and all the other forms in which ***ITS*** weight exists, such as the celestial bodies that exist with in ***ITS*** nothingness. All this enables us to begin to understand how ***IT*** took ***ITS*** once singular weight, and become everything that now exists as ***ITS*** weight, and also how ***IT*** takes this fragmented weight back to a singular point again using what we refer to as black holes.

So to ***IT*** I say: "Thank you , for letting me understand you a little better, as ***YOU*** exist as life; as your nothingness and as your weight that exists with in your body, that exists as this cold, clear, empty universe, where you reshape your weight in, so that I could say thank you again and again."

I end this piece by saying that from what we were back in the caves to what we are now, so that I could write and you could read about ***IT***, is a very interesting and awe-inspiring show of ***ITS*** reshaping ***ITS*** weight within ***ITS*** nothingness.

ITS weight as a law

Here is a situation concerning which I am not sure if there is a scientific law already formulated. Let me present ***IT*** this way: There should be a law that applies to ***ITS*** weight, and to ***ITS*** weight only, which should address the fact that ***ITS*** weight will always be in the process of changing, since ***ITS*** weight is always wanting to reunite again as one singular weight. And as we have seen that ***IT*** is a duality, there should be a converse complementary law addressing the fact that ***ITS*** huge nothingness will always stay constant, for as ***ITS*** nothingness, ***IT*** will never change ***ITSELF***. On the other hand, spiritually speaking, ***ITS*** divine nothingness, which also includes life and ***ITS*** many attributes, such as love, hate, and passion, my gut feeling is that ***IT*** is here where these attributes reside.

Why nothing lasts forever

When we refer to nothing lasting forever we are referring to ***ITS*** weight, which is what is never lasting as one entity. For example, we know that our Sun will not last forever as ***ITS*** weight, and when we say that a human relationship will not last forever, we are really referring to how we exist as ***ITS*** weight, not as ***ITS*** constant nothingness. Another thing that we should understand is that ***IT***, as ***ITSELF***, has no distance, in the sense that we consider between someone and someone else. You will become aware of this if you realize that to speak about ***IT*** in terms of distance there would have to be at least one more god so that we could measure the distance from one god to the other, for we cannot apply distance to just ***ITS*** nothingness, for how can we measure something that is made of a nothingness?

Transferring ITS weight

When we refer to ***ITS*** energy being transferred, what is really being transferred is ***ITS*** weight, as heated energy, for ***ITS*** weight is always seeking to change into a larger or smaller weight configuration. Not so ***ITS*** nothingness, which remains constant.

Maybe you might see ***IT*** better this way: Even before we leave our beds to start our daily activities, as we sleep our bodies are continually transferring energy, for our body contains heat as weight, and if the energy is not replaced as food, for what ever we consume contains heat as weight, we will notice a drop in our weight, and one reason for this that we exist as a part of ***ITS*** heated weight. We will notice this heat when we sweat. This heat is equivalent to weight. This has been confirmed by athletes, who can lose pounds through performing an activity that requires a lot of energy, such as boxing.

⌘~~~~~~~~~~~~~~~⌘⌘~~~~~~~~~~~~~~~⌘

*** ***When we fight wars we are conquering and destroying (reshaping) ITS weight.*** ***

❀~~~~~~~~~~~~~~~❀❀~~~~~~~~~~~~~~❀❀~~~~~~~~~~~~~~~❀

But let me get back to ***ITS*** weight. As we leave sleep we are going to be continually transferring energy. Now here is the thing, that we, like just about everything else that is transferring energy, is just transferring ***ITS*** weight, for ***ITS*** nothingness is not taking part in this, and the reason is that ***ITS*** nothingness remains constant, for ***IT*** is ***ITS*** heated weight that is continuously changing as ***IT*** reshapes to a larger or lesser weight.

⌘~~~~~~~~~~~~~~⌘~~~~~~~~~~~~~~⌘~~~~~~~~~~~~~⌘

******It is our purpose in life to move ITS heated-weight******

❀~~~~~~~~~~~~~~~~~~~~~~~~~~~~~~~~~~~~~~❀

Change and constancy

Here is an idea for those readers that have heard a saying that goes like this: "What you are made of never changes; who you are is always changing". When I heard this from a friend, I asked myself how did someone know that we are made from that part of ***IT*** that never changes, as the constant nothingness that we exist as, and every thing else that may exist as the nothingness that atoms exist as, and how did someone know that who we are as ***ITS*** weight (matter) is always changing? For these reasons, I am going to rephrase the saying "what you are made of never changes, what you are, is always changing" to "**we exist as *ITS* never changing nothingness, where we can participate, with *ITS* constantly changing weight".**

IT does not sleep

Here is an idea that came to me as I went to sleep at night: I know that I do need to rest and sleep, yet I know that ***IT*** does not sleep, because if ***IT*** did who would run the Universe or rather, ***ITS*** constantly changing weight? In observing ***ITS*** ways I have come to the conclusion that, as ***ITS*** nothingness, ***IT*** does not need to rest because ***IT*** does not exist as something.

You might see ***IT*** better this way: We need rest because we exist as something, that is, as ***ITS*** weight, yet what keeps us alive is ***ITS*** nothingness, which is why we, as life, don't close down to rest, the process of life does not stop while we are asleep. However, it is ***ITS*** weight as us that needs to slow down in order to reshape. You may see this better if you consider a baby, which needs rest so as to accumulate more of ***ITS*** weight in order to grow, as ***ITS*** weight.

Here is another interesting thing: Consider ***ITS*** weight as the atom, which is the manner in which ***ITS*** weight exists so as to keep ***ITS*** fragmentation of ***ITS*** weight in one place. Atoms have a nucleus which must have at least one proton, and naturally, this proton which came from ***ITS*** ever changing weight that is now at rest and as a fraction of ***ITS*** total heated weight it also has to exist within ***ITS*** never-changing nothingness.

The proton is at rest because of the spin that ***IT*** has, (by the way, electrons also have spin), and I say that this weight is at rest due to ***ITS*** spin, because as I have also watched ***ITS*** weight at rest as the Sun and Moon, which are also resting with in ***ITS*** nothingness. This is due to the spin that the Sun and Moon have.

So I end this by saying: “Thank you (***IT***) that you do not sleep, for who would watch your weight as you slept?”

It is ***ITS*** weight that is at rest in order to stay in one location, just like the proton in the atom, and the Sun and the planets in our Solar System, and the stars throughout the Universe.

MAXX-SPEED, a speed faster than 186,000 mps

I know that by saying that there is a speed faster than 186,000 mps, I will hear from everyone that this cannot be. But I am only using this number as an example and as a question to the professionals who would know more.

First of all, the speed of 186,000 mps as light did not exist before the Big Bang, and I do think there is a faster speed. I will agree that 186,000 mps exists in the absence of matter, and we say that light travels at this speed. The human mind states this, and it is based on our concept of distance and our mechanical time system, and we make all these calculations based on the rotation of our planet. It is a convenience for us to be able to make these calculations as such.

But time did not exist before the Big Bang. Whatever the speed is that light uses to catch a ride, it exists as ***IT***. Let us reason together: The speed of 186,000 mps could exist in the absence of matter. Matter, as the weight of ***IT***, could be what is slowing down the 186,000 plus speed. If light behaves as both a wave and as a particle that generates heat, which has weight, then light has mass. How can the speed of light be the fastest? What about the speed that exists without light in pure nothingness? Wouldn't it be faster without the slight mass that light carries as a particle (photon)? Perhaps this speed does not matter, but if it exists, it is also ***IT***. And ***IT*** exists as one that has an area where the temperature is below zero and is made of nothingness. It also has concentrated weight that has heat. This is easy to see, but I bump into a wall when I think about ***IT*** being this other speed that we have not been able to perceive, let alone measure, so this query becomes neither religion nor science. Yet the extreme emptiness of space remains, and it is ***IT***. To most, ***IT*** does not exist; but there ***IT*** is. Dark Matter is not really dark, and it is not really matter; it is clear and cold, and likely is moving faster than 186,000 mps.

In addition, speed acts as a duality to ***ITS*** weight. One, ***IT*** moves weight outwards as energy from the Sun and other stars, and two, ***IT*** holds weight in as the electrons. And maybe, if speed also exists independently in ***ITS*** cold nothingness, instead of a duality, speed could be a trilogy, for we have: one, ***ITS*** weight; two, ***ITS*** speed; and three ***ITS*** nothingness where ***MAXX-SPEED*** exists totally and faster than 186,000 mps. But to be sure, we would have to know the speed that already exists in the atoms of matter as well as the speed that ***IT*** uses to transfer weight outwards from the Sun and other stars because what we do know is that if speed did not exist, then the Universe as ***IT*** now exists as 5% matter could not exist.

The speed of light

When talking about the electron moving at the speed of light, which science has defined as 186,000 miles per second (abbreviated here as 186,000 mps), we must remember that this phrase and this number are human statements that only apply to that which exists after the Big Bang. And more important is that this speed of 186,000 mps is based on our human mechanical time system.

I say *we* must remember because I have had to train myself in my ways of thinking so that I could begin to understand ***IT*** better as to the way ***IT*** really is, and my hope is you are also trying to see from new perspectives. Let me explain how by using this 186,000 mps speed as ***IT***. There was a moment when I was used to seeing 186,000 mps as what we, as humans, call the speed of light, but now that I know that everything exists as pure energy or ***IT*** as omnipresent I have focused my thinking on the duality of ***IT*** that includes not only the speed of light but the qualities of this speed before the Big Bang. We can see speed before the Big Bang by eliminating from our image of the Universe all the things that we now see as existing, our Sun for example, which is where we get the use of the word light, and our planet too, which is what gives us miles per second as time. The Sun and the planets that exist as ***ITS*** weight (matter) did not exist before the Big Bang as matter, yet ***ITS*** weight did exist and as a very dense energy. If it were not for ***ITS*** weight, humans would never have come to exist, and if we did not exist, what we have come to know of ***IT*** as God, omnipresent, or as pure energy, or any of the other words that we use to confirm ***ITS*** existence, would not exist either, and it is here that we can see how ***IT*** as speed existed before the Big Bang.

We know that the more weight this speed has, the slower ***IT*** moves as ***ITS*** weight. So that when one looks for the speed ***IT*** existed as before the Big Bang, one will not find ***IT*** as the Sun or planets slowing speed down, for ***IT*** had not yet reshaped ***ITS*** weight into those things. Still, however, the weight is there, as is this speed, for speed is not something that we produced, but rather it is something that ***IT*** existed as before the Big Bang, as energy, because anything that can go that fast has to have energy.

So before the Big Bang, speed, as energy, existed, but not as matter, which leads us to see this speed in cold nothingness. I say cold because speed is the reverse of matter, and this cold, clear nothingness that ***IT*** exists as is the reverse of ***ITS*** heated weight, as the duality that ***IT*** exists as.

I can see that if it were not for ***ITS*** weight we could not exist as having weight, and that if it were not for ***ITS*** speed, as ***ITS*** duality, we could not have come into existence after the Big Bang. We needed this speed in order to exist as the electrons that make our atoms and all matter possible.

And I now see people like Charles Darwin, who showed us ***ITS*** reshaping as evolution, as change and as something, a person, Darwin, who made it easier for humans to understand ***IT*** better. People like Charles Darwin have shown me ways in which ***IT*** evolves as ***IT*** reshapes. I know that Darwin, like Einstein, could not have existed if it were not for the weight and speed that ***IT*** used to make ***ITSELF*** as a person named Darwin, so that I could see ***IT*** better as ***IT*** was evolving.

I now see people like Einstein as a way ***IT*** existed in order to provide us understanding as to ***ITS*** existence as energy. ***IT*** permitted Einstein to understand ***ITS*** weight as energy and the formula $E=mc^2$. I feel that in opening up the databank on my webpage, we may be able to find more on ***ITS*** energy as weight and speed as readers share what they know, or as the people with more information print their journals for the rest of us to read.

There is much to understand about ***IT*** as weight, speed, and the many other characteristics that ***IT*** exists as.

I am also sure that people in the non-scientific professions like me , when looking at and relating to ***IT***, will similarly make our understanding of ***IT*** clearer for ***IT*** is also made of clearness; for even ***ITS*** speed is clear.

Finally, as for speed, we might have to adjust our way of seeing change, for ***IT*** may pick up the pace of or the speed of our daily changes. And now that ***IT*** has reshaped into the speed of our ever growing technology, ***IT*** might be better if we are not on the assembly line because production is going to be much faster!

Allow me to begin this section by explaining why I will be referring to ***MAXX-SPEED*** as 200,000 mps. One reason is because as 200,000 mps is an easy number for you, the reader, to remember and it is an easy number for me to use in other areas in this book where I have to refer to ***IT*** as being faster than 186,000 mps.

For example, I would like to use this number to describe the way ***IT*** might exist when we remove the weight that light has as a particle (a photon). Scientists have calculated the normal speed of light as 186,000 mps, but relieved from this weight, light can travel faster. I will use 200,000 mps as a convenient number and call it ***MAXX-SPEED*****.** Conversely, when ***IT*** uses more of ***ITS*** weight, for example, to shape into electrons, ***IT*** exists at a speed slower than 186,000 mps, which is what we refer to when we say that electrons travel near (that is, slower) than the speed of light, meaning that electrons must travel at less than 186,000 mps. The opposite to this would be, as I said before, that when the weight of a light particle (photon) is removed (for light is something and if something does exist, it must have a portion of ***ITS*** weight attached to it), it will be able to move at a speed faster than 186,000 mps. To this faster speed I awarded the number 200mps for the sake of convenience and simplicity throughout this book.

Speed of 186,000 mps as a human concept

The phrase that refers to nothing going faster than the speed of light is true only when this speed is in the presence of matter. We have to remember that just before the Big Bang we could not use this statement of nothing going faster than the speed of light because at that stage of ***ITS*** existence there was no light, not from any stars, at least, for the speed of light to exist.

You can understand then, this ***MAXX-SPEED*** can exist as something independent of light and matter, for the speed of light, as we refer to it, is a by-product of the Big Bang, and ***ITS MAXX-SPEED*** cannot be created or destroyed. ***MAXX-SPEED*** could be the bulk of ***IT***, as the pure energy that now exists, as the distance that ***IT*** spans, as the nothingness that this Universe exists as.

⌘~~~~~~~~~~~~~~~~~~~~~~~~~~~~~~~~~⌘

*** ***We cannot exist without speed.*** ***

❀~~~~~~~~~~~~~~~~~~~~~~~~~❀

One more way to see IT

Here is one more way to see ***IT***: Imagine that ***IT*** is one huge ball of weight, something like what we have been talking about when we refer to how ***IT*** existed at the moment of the Big Bang, and ***IT*** took ***ITS*** weight and chopped it up into discrete, quantifiable bits. And then, using these bits, ***IT*** put together the more than 92 naturally occurring elements in the Periodic Table. With these fragmented bits or elements, ***IT*** put together what we see as matter in this Universe. I am very thankful to ***IT*** for using ***ITS*** weight as all the elements that I, you, and everyone else are made from, including our food, clothing, and housing, so that we could understand that ***IT*** does exist, and that ***IT*** exists as everything that we can understand as being in this place called omnipresent. I should also say that we do know that ***IT*** exists as the pure energy that now exists as this Universe, which is where ***IT*** exists as ***ITS*** duality, that is, as both the empty nothingness and as the matter that exists as ***ITS*** weight, or to state ***IT*** in other words: as atoms that exist with in ***ITS*** nothingness.

IT is ***ITS*** weight that is reshapeable. When ***IT*** is done by us, for instance, when we use ***ITS*** weight to build things, ***IT*** does give us resistance, for example, when we use ***ITS*** weight to build concrete buildings, planes, cars, or boats.

However, when ***IT*** uses ***ITS*** weight to reshape, ***ITS*** weight offers no resistance to ***ITSELF***, but rather will just fall into place, such as in those things that ***IT*** naturally puts together. Some examples of this would be the weight of our human body, plants, fish, birds, or as in every thing that is made by nature ***(IT)***:

1- Here is something that we should remember: When we use the word black, we are referring to the color black, and when we use the word darkness, we are referring to the absence of light, and that the things that are black, for example some minerals, came from the elements that came into existence after the Big Bang. This could mean that colors are a property of ***ITS*** weight, not ***ITS*** nothingness.

So that when we say that in light all the colors of the spectrum exist, they may come from the colors that exist as ***ITS*** heated weight, in the form of light. Let me take a moment to clarify that what is coming out of the Sun is not light. What we perceive as light is ***ITS*** heat that is catching a ride as ***IT*** transfers ***ITS*** heated weight to other parts of ***ITS*** heated weight.

Look at ***IT*** this way: When we turn on a light bulb we see the light in the surrounding area, let's say "room", as the glare. This glare is only about 4% of the energy put out by the light bulb.

When our planet is facing the Sun, we see light as the glare that exists near the sun. Probably someone reading this will remember that arriving light has been recorded as being both a wave and particle. I agree, for what we measure as light as this wave and particle is ***ITS*** speed as a wave, and ***ITS*** weight as the weight of the heated particle, but not as light, or brightness, for the brightness from the Sun stays near the Sun, the same way the brightness from a bulb stays near the bulb.

Maybe you will understand ***IT*** better this way: When we place our hands near a light bulb what we will feel is not ***ITS*** light, ***IT*** is ***ITS*** heat, for light happens when both aspects of ***ITS*** duality come together as one.

This is the same for our Sun. What we see from a distance is the glare from ***ITS*** duality making contact, and what we feel is ***ITS*** heated weight, that is hotter as we get closer to ***ITS*** origin (the millions of degrees that exist as ***ITS*** heat as our Sun), which involves the same process that we experience when we get close to a 200 watt light bulb.

Here is one more way to see colors as ***ITS*** weight as light: In order to see ***ITS*** colors we do need to see something, and if this something does exist, then it is made from ***ITS*** heated weight. But you will see the opposite of this when you look at ***ITS*** nothingness, where nothing exists, not even colors, wherefore nothing can be seen, because if this nothingness that now exists as this Universe had colors, ***IT*** would make ***IT*** harder for us to see through ***ITS*** way of existing as a kind of a cold clear nothingness.

2- As to how powerful ***ITS*** weight as energy may be, just think of all the electrical things that exist, and remember that electricity involves using just a tiny fraction of ***ITS*** weight as the energy that we know as electricity. ***IT*** is so powerful and dangerous that we scrupulously avoid touching ***ITS*** weight as electricity. An electric shock of sufficient voltage can kill you!

3- From black holes to the beginning of the Big Bang

First of all, black holes are where ***IT*** begins to bring back ***ITS*** fragmented heated weight that exists within this Universe as matter, so as to prepare a concentrated portion of ***ITS*** weight, but not as matter. When I say "not as matter", what I mean is not as in matter that has electrons, because what is being pulled inwards is ***ITS*** heated weight that existed as the heat inside atoms.

We should remember that ***ITS*** heated weight is always seeking to return to ***ITS*** singularity as weight, as ***IT*** was before the Big Bang.

As for the matter that is being pushed outwards into the now existing Universe (***ITS*** nothingness), which is becoming infinitely smaller as ***IT*** spreads out, ***IT*** could eventually be pulled back to ***ITS*** original singularity as weight, as waves that still contain the missing weight from the original dense weight that existed as a singularity, before the Big Bang occurred. This is one possibility.

The second possibility may be that the dense matter that now exists within this Universe as ***ITS*** weight, which is where ***IT*** keeps ***ITS*** weight within ***ITSELF***, could be decomposed, fragmenting in to smaller bits, making this weight become infinitely smaller, so that

IT could reshape to become ***ITS*** stage 3, in which ***IT*** has all ***ITS*** weight and nothingness existing as one uniform self, as weight and as temperature.

My personal feeling is that ***IT*** will continue to return to ***ITS*** stage 1, when ***IT*** had all ***ITS*** weight in one place and later fragmented this weight (at the moment of the Big Bang) so as to reshape into atoms (matter). I say this because if ***IT*** does have a stage #3; at this stage ***IT*** is neutral, where nothing is basically happening to ***ITSELF***, or not as much is happening as in ***ITS*** stages 1 and 2; and as stage #2 we have already confirmed that ***IT*** can bring back ***ITS*** heated weight through what we call black holes, which are really clear concentrated hot spots.

And one reason why we cannot see black holes is because this weight exists as clear weight, you might understand ***IT*** better this way, since this weight does not have the colors that come from matter, as minerals do.

And one more reason why we cannot see black holes is because whatever light hits the black hole will not be reflected, for in this case light is not hitting the black hole as is the case with a solid material object. Rather, it is being pulled inwards as weight, and will be totally taken inwards because of the gravitational pull that the black hole has as weight, (as ***ITS*** mass), so that the weight that light has will now be added as more heated weight to the clear hot spot that now exists somewhere inside of ***ITSELF*** as ***ITS*** nothingness.

All matter has electrons and when black holes pull in matter, these electrons will be stripped away, and this is so because since the electrons have weight attached to them this weight too will have to return to ***ITS*** original heated weight that now exists as a black hole, as clear heated weight. As the electrons are stripped of the weight they carry, then the speed of 186,000 mps, will become faster, since ***IT*** does now not have the particles of weight that were traveling as electrons, and that now that this faster speed can exist as ***MAXX-SPEED*** because it has no weight, it cannot be pulled inwards in to the center of the black hole as gravity, so that this ***MAXX-SPEED*** can return to ***ITS*** weightless area that ***IT*** exists as the nothingness that exist as this empty Universe, which is where this

MAXX-SPEED can naturally exist as ***MAXX-SPEED*** or as ***ITS*** nothingness, because ***MAXX-SPEED*** is ***ITS*** nothingness as this existing empty Universe, that is: ***ITS*** clear cold inner and outer shell, as ***ITS*** body.

The reason why not even light can escape a black hole is because since light can behave as both wave and particle, these photons will be pulled inwards to become incorporated or pulled in, or not being able to escape, for the weight that exists in light has to stay inside the black hole with the rest of the weight that is already inside as this weight continues to increase in size. When this weight that light has when traveling at 186,000 mps is pulled inwards, it will become ***MAXX-SPEED*** again, and become the nothingness that it naturally exist as, because this high speed cannot be created or destroyed.

Our perception of mass is necessary if we are to see ***IT*** as something that does indeed exist. And this way of seeing mass was needed for physics, in order to understand this pure energy that exists out there. We also need to see the mass of objects in order to confirm that something does exist. Though I know that weight and mass are two different concepts, and that weight varies in direct proportion to the amount of gravitational force present in a certain area, while mass remains constant, I use the terms interchangeably. Now there is no problem with seeing ***ITS*** weight this way, but if we want to understand ***IT*** better as ***ITS*** weight (mass) ***IT*** helps if we see ***IT*** as weight also, because when we focus on ***IT*** as mass, we will be seeing ***IT*** more as matter. However, ***IT*** began existing as matter only after the Big Bang, but we should keep in mind that ***ITS*** weight also existed before the Big Bang, as weight, but not as the matter that we now understand is made up of atoms. This way, when we focus on ***IT*** as ***ITS*** weight only, we will see ***IT*** from a different perspective as how ***IT*** exists as ***ITS*** weight. In the same way when we focus on ***ITS*** nothingness, we will also understand ***IT*** better from yet a different perspective.

Here is one more good question for a reader that might know the answer, and the question is this: We know that ***IT*** exists as duality of hot and cold. For instance, as in the extraordinary heat that existed at the moment of the Big Bang, but ***IT*** also exists as the freezing cold in the vast areas of nothingness that exist in this Universe, which is ***ITS*** invisible inner and outer clear shell (body). So the question is: How can a freezing cold nothingness exist in the measurable area that this Universe exists as, where a concentrated heat can also exist simultaneously?

Let me try and present this a different way: ***IT***, as pure energy, exists as a duality of temperatures, and these two are properties of this pure energy, which as such, cannot be created or destroyed. Yet the dimensions of ***ITS*** coldness are huge, existing as a nothingness, while ***ITS*** heat exists more in a concentrated area that has weight, but is not solid, and can fragment into finite portions, such as photons.

I know this is a strange question, but we should remember that this God that does exist is in all places as the same moment, which means that ***IT*** exists as this vast freezing cold Universe, and ***IT*** also exists as the heat that exists in every atom, even in those atoms that you and I are made from.

Here is one more way to see this: Whoever or whatever this Being is, ***IT*** is in all places as the same moment, and ***ITS*** basic components (or way of existing) are ***ITS*** coldness as ***ITS*** body, and ***ITS*** weight as heat, that exists inside ***ITS*** coldness.

So here is something to ponder until some one makes ***IT*** clearer: If ***IT*** exists as both cold and heat, as this Universe now exists, maybe as a possibility ***IT*** could combine both these temperatures, which would mean that both ***ITS*** nothingness and ***ITS*** somethingness as heat can also exist as one equal temperature and weight. If this is so, then how did ***IT*** divide ***ITSELF*** into these two extremes, becoming one huge freezing cold nothingness (the Universe) and also one extreme local source of heat, such as the heat that existed at the moment of the Big Bang? My feelings are that ***IT*** exists as one, as in the descriptive religious phrase "one God", one Supreme Being, or one pure energy, where both of these two extremes exist in ***ITSELF*** as one. In fact this is what we are now seeing in this Universe, which is ***ITS*** body, being made up of the freezing coldness, where ***IT*** keeps ***ITS*** extreme heated weight inside of ***ITSELF*** with out blending these two extreme temperatures. Let me explain this in a different way: The heat that existed at the moment of the Big Bang was not destroyed, but fragmented into the heat that now exists in atoms, which resides within ***ITS*** freezing coldness, and which do not blend or mix with ***ITS*** cold nothingness to become an even temperature, for when we observe this Universe, we can see that ***ITS*** heat is still there, and ***ITS*** coldness is still there, as two different properties.

Now, since I am no pro in temperatures, let me share what I see as an inexperienced observer: When ***IT*** fragmented ***ITS*** weight into protons, ***IT*** kept this heat separated from mixing with the outside freezing temperature by placing electrons as separators, and ***IT*** did this by using ***ITS*** ways of existing as a oneness; on the one hand, as electrons, which have ***ITS*** speeded nothingness along with a little of ***ITS*** weight, in order to keep the heat inside from mixing with the coldness that exists outside.

Here's another question for the pros: Does anyone know what temperatures exist inside the atom? Scientists say that atoms are mostly composed of empty space. What is the temperature in this empty space? What is the temperature of the area outside the atom's nucleus? Now, since ***IT*** exists as these two extreme temperatures, i.e., hot and cold, this may lead us to understand what we call positive and negative. I say this because, as pure energy, we refer to protons ***(ITS*** heat) as the positive, and since this which we refer to as positive was present as the dense matter (***ITS*** weight) at the moment of the Big Bang, then logically we would have to refer to ***ITS*** cold, clear, freezing cold nothingness as the negative: ***ITS*** opposite.

It could also be that these two extreme temperatures as ***ITSELF*** cannot be merged. Cold stops things from happening, at least as ***ITS*** weight. And heat as ***ITS*** reverse is always making things happen, so that as ***ITS*** coldness nothing is really happening, for this is where ***IT*** exists as a constant in a state of rest, without any movement, and ***IT*** is in ***ITS*** heat or hotness that something is happening, as change.

⌘~~~~~~~~~~~~~~~~⌘⌘~~~~~~~~~~~~~~~~⌘

******God, as pure energy, exists as a freezing cold body, where IT keeps ITS heat inside ******

❁~~~~~~~~~~~~~~~~~~~~~~~~~~~~~❁

Sometimes we have to take a different approach in order to understand things. As an example, we are used to seeing things getting hotter and colder. Yet ***IT*** exists as these two extremes at the same time, in different places, as ***ITS*** way of existing, so that if we take these two extremes and apply our understanding of ***IT*** as pure energy, where nothing is really created or destroyed, then these two extremes cannot be created or destroyed either. If this is so, then we will need to adjust our way of thinking concerning how things become hotter or colder. For ***ITS*** two extreme temperatures cannot be created or destroyed, so that we might have to accept that these two extreme temperatures can exist, independently, as properties of this pure energy.

Here is one more way that will help you to understand ***IT***, as pure energy: The pure energy that is ***IT*** is composed of 2 ways of existing: one, as the cold, clear nothingness that now exists as this Universe and two, as heat. We know about the heated dense matter that existed just before the Big Bang, which is now the same heat that exists inside everything that we call matter. Now, here is something that is very important: The heat that we exist as, the atoms (protons) that make up our bodies also have to exist inside ***ITS*** nothingness, in the same way a black hole exists as ***ITS*** heat, which the black hole acquired when ***IT*** pulled in the surrounding matter, after which this now heated black hole continues to exist independently within ***ITS*** nothingness.

Remember that a black hole was once a supergiant or a nova that existed as ***ITS*** heated weight, so all black holes have to exist within ***ITS*** nothingness. For nothing that has weight can exist outside of ***ITS*** nothingness.

⌘~~~~~~~~~~~~~~~~⌘⌘~~~~~~~~~~~~~~~~⌘

*** ***We are made in ITS own image: our bodies are made up of atoms, and as such have the property of heat, and at the same time we also contain the nothingness that exists inside each atom.*** ***

۞~~~~~~~~~~~~~~~~~~~~~~~~~~~~~~~~~~۞

And when we refer to black holes having the power of pulling in matter, where not even light can escape, I believe that ***IT*** is not a case of *light* that cannot escape. What is really taking place is that ***IT*** is ***ITS*** weight that is being pulled inwards, because the power that weight has as a black holes' gravitational pull, is that as gravity, ***ITS*** weight has the ability of pulling in more of the surrounding weight, for weight is just really looking to reunite itself as heated weight back to when it existed as a singularity.

So that as matter is being pulled into a black hole, ***IT*** will be pulled to ***ITS*** center and since matter is composed of atoms that have electrons and these electrons also have weight attached to them, this weight will be stripped from the electrons making them lighter which will allow them to return to ***MAXX-SPEED***, and at ***MAXX-SPEED IT*** is capable of returning to ***ITS*** original place, which is back to ***ITS*** cold, clear nothingness. This is even more so because this speed is made from a coldness, this is the coldness that exists within ***ITSELF*** as ***ITS*** housing, for we should remember that ***IT*** was ***ITS*** clear heated weight that ***IT*** utilized to form matter within ***ITS*** cold, clear speeded nothingness.

Let me show you a different way of looking at this: Since ***ITS*** heated weight is always pulling inwards, the opposite of this is ***ITS*** cold nothingness, which is looking to be on the outside of ***ITS*** heat, but on the inside as ***ITSELF***.

Let me try ***IT*** one more way: When the heated weight that exists in light is removed, then light can travel faster than 186,000 mps, just returning to ***ITS*** original faster speed (this is what I call "***MAXX-SPEED***").

Consider this: When light was moving at 186,000 mps, it was carrying ***ITS*** heated weight to be transferred elsewhere as weight. Let us pose a hypothetical example using our own Sun and a black hole somewhere close by. In this case, in order to return ***ITS*** heated weight from the Sun (hydrogen and helium atoms), as they are pulled into the black hole the atoms' electrons are stripped of their weight, so that it can now be returned to a central singular point again, as the density that is related to black holes, without the emptiness that existed inside the hydrogen and helium atoms that were once part of the Sun.

Let me add something here that I hope will make sense to you, the reader, when I refer to ***ITS*** heated weight being clear, as I have said before, I mean transparent: If ***ITS*** heated weight were not clear in ***ITS*** composition, we would not be able to see through the heat that exists as protons and neutrons that are inside oxygen atoms for example; and if ***ITS*** heated weight were not clear, then for example, when our astronauts look back at our planet, they would not be able to see through the tons of heat that the light of the Sun carries to our planet.

Here is yet one more way to understand this: What we have been calling black holes are really ***ITS*** clear heated weight, without ***ITS*** speed (the electrons) and the emptiness that matter exists as, for in the case of a black hole this clear heated weight has to exist as ***ITS*** singular heated weight only, which will be surrounded by ***ITS*** cold, clear nothingness.

So black holes are there to remove the weight that exists as electrons, that ***IT*** placed there in order to maintain the separation of ***ITS*** heated weight, so that atoms, as matter, could exist, and by removing the weight that was also part of the electron ***IT*** can bring this weight back to a singular point again in preparation for bringing back all the weight that ***IT*** fragmented at the moment of the Big Bang, so that ***IT*** could regroup these heavier portions that ***IT*** compressed as black holes. Now, all ***IT*** has to do is use the gravitational pull that ***ITS*** heat has to bring (or eat up or pull in) all the other black holes, so as to form just one concentrated weight again. When that happens, ***IT*** is **back to the beginning of what we have been calling the moment of the Big Bang,** and may choose to repeat ***ITSELF*** again as something new and different or ***IT*** may choose to just exist as ***ITSELF***, as the same never ending omnipresent moment.

Let me clarify that the reason why I say that black holes are really **clear heated weight** is because black holes are pulling in matter that existed as protons, neutrons, and electrons, that themselves exist as heat inside the atoms. After being sucked in, this heat is now inside the black hole, and this is the reason why we cannot see black holes.

⌘~~~~~~~~~~~~~~~~⌘⌘~~~~~~~~~~~~~~~~⌘

****** The heat we feel from light is a fragment of ITS clear heated weight ******

❁~~~~~~~~~~~~~~~~~~~~~~~~~~~~~~~~❁

I regret that I again have to be repetitious, but do you remember a while back when we talked about the clarity that ***IT*** also exist as, which we can confirm just by looking straight in front of us? And do you also remember that the matter that exists outside of us, such as oxygen atoms in the air, that has all the elements that ***IT*** exists as, as matter, as the atoms' electrons, and their heated protons and neutrons, and also the clear emptiness that exists inside oxygen atoms, that are so clear that they are invisible to our eyes? Well, black holes are made from this same clear heated weight and this is why right now we cannot visually see any black holes, for they are truly **clear, concentrated hot spots.**

Unfortunately since these black holes are made of a concentrated clear heat, we cannot get too close to them, because since this heated area is made from ***ITS*** heat which is seeking to pull any surrounding heated weight back to a singular point, we or anything that has ***ITS*** heated weight will be pulled in. The only thing that I can think of that can help in detecting the presence of these black holes (clear dense hot spots) is a huge temperature thermometer that can detect heat from far enough away, where the actual thermometer will not be pulled in too!

Let me end this section, by saying something that I have felt during my writing on this and other subjects, and it is that I find that the information above only helps me to understand ***IT*** better in what concerns the ways that ***ITS*** weight exists and performs, within ***ITSELF***, as my outside existence, for none of the above helps me to feel better physically and mentally. What I have found that does make me feel better is to continuously connect with ***IT*** inside of me. But I am grateful for the information above, because ***IT*** was information that I too did not have before I started writing this book, as to who ***IT*** is, and how ***IT*** exists, and I feel that some readers will be able to use the information that I have put into this book, and they will, as a spin off, come up with something that I could not do or was not physically here to do, in the way they where destined to, for I know that all my readers have information on ***IT*** that I do not yet have or know. So again, if you have something on ***IT*** that I have not mentioned and would like to share ***IT*** with others in my web page data bank, please send ***IT*** to me as an e-mail or to my P. O. Box at 9944, Carolina,P.R.00988-9944 as postal mail.

And while I am on this subject of black holes, let me share with you some other bits of information that will help you understand more of how ***IT*** exists as black holes.

IT made black holes by first putting ***ITSELF*** together as ***ITS*** weight, as the matter that first had to exist as supergiant stars or supernovas, so it looks like ***IT*** is here where ***IT*** starts to remove some of ***ITS*** emptiness that exists as matter (the atoms that still exist inside stars), in order to begin to become denser heated weight again, for as denser heated weight ***IT*** can continue to pull back ***ITS*** smaller fragments of loose weight that exist outside the black hole areas.

Furthermore, I assume that after ***IT*** has pulled in all of the surrounding weight that ***IT*** exists as in the form of galaxies, the last galaxy which is now a concentrated heated area that shall be left to ***ITSELF***, after ***IT*** has consumed all of the matter that existed as ***ITS*** surroundings, (what we call galaxies), the now one singular hot spot (black hole), will pull or be pulled by the next hot spot (black holes that may exist nearest to this clear hot spot), and ***IT*** will eventually bring all of ***ITS*** individually separated heated weight that exists in different parts of ***ITSELF***, for all of these concentrated hot spots will continue pulling themselves together, into one big black hole, eating up all the other black holes. Let me take a moment here to clarify that black holes do not eat themselves up, for what is happening is that ***IT*** is pulling in ***ITS*** outside weight, so as to reshape ***ITSELF*** back to being a more concentrated dense heated weight.

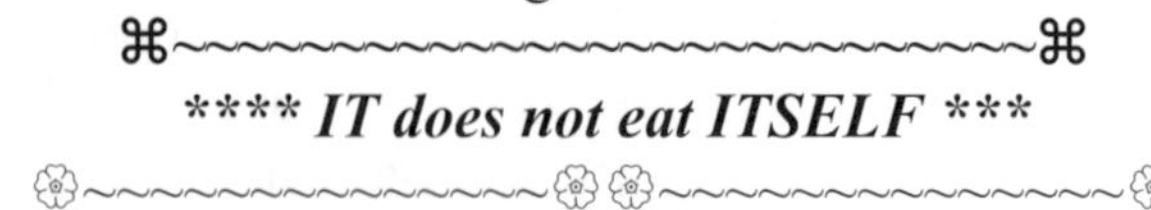

**** *IT does not eat ITSELF* ***

I think ***IT*** is better to understand if we look at ***IT*** this way: What ***IT*** is doing as black holes is that ***IT*** is pulling together all of the different concentrated hot spots that exist in different parts of ***ITSELF*** (the existing Universe).

Maybe you will understand this better this way: ***ITS*** heated weight that now exists as different parts of ***ITSELF*** in the form of galaxies, will be pulled back again to a central point within ***ITS*** nothingness, just the same way ***IT*** existed when ***IT*** had all ***ITS*** heated weight in one area, which is what we call the dense matter that existed at the moment of the Big Bang. Or look at the Big Bang in reverse: the dense heated weight that now exists inside a black hole is the same heated weight that existed at the moment of the Big Bang. This has to be so because of what we know as the Law of Conservation of Energy. And just as the meaning of the word omnipresence, nothing is ever created or destroyed, just transformed or reshaped.

So that as two black holes reunite, by one consuming the other, the process will continue consuming other black holes, because the first black hole that consumed the other one is now bigger, and in being bigger, with ***ITS*** extra pulling power as gravity, ***IT*** will more easily pull in other smaller black holes.

Here is one more way to see this, but first we should remember that, all matter as heated weight that exists with in ***ITSELF*** as ***ITS*** nothingness (the Universe) has to stay within ***ITSELF***, as ***ITS*** nothingness: When ***IT*** threw out ***ITS*** heated weight at the moment of the Big Bang, this weight fragmented into what we now see as the matter that now exists within ***ITS*** nothingness (this Universe) so that this fragmented weight as matter, is being reunited back to a central singular point, just the same way ***ITS*** heated weight existed when ***IT*** existed as the very dense heated matter at the moment of the Big Bang.

So you see, ***IT*** brings back all ***ITS*** once singular weight using ***ITS*** weight as gravity, which is what ***IT*** reshaped ***ITS*** heated weight into as supernovas, and afterwards as black holes, so that ***IT*** could continue to build up more pulling power as the gravity that ***ITS*** heated weight has, so that each individual black hole could pull other black holes together to form once again one total black hole, which would be the same as what existed at the moment of the Big Bang.

It is evident then that clear hot spots, (black holes) is what ***IT*** reshapes into in order to reunite ***ITS*** heated weight again, within

ITS nothingness, in order to return to how ***IT*** existed as that (same) moment again, that we call the moment of the Big Bang. Now it is possible to see that maybe this is the way that ***IT*** restarts again.

Here I would like to say (and I am only making a reasonable assumption), that every time ***IT*** restarts ***ITSELF*** as a different Big Bang, using the same very heated dense weight that ***IT*** exists as within ***ITSELF***, something new may happen, because of ***ITS*** weight being distributed differently within ***ITSELF***. This is one possibility. Another possibility may be that now that ***IT*** knows what ***IT*** could eventually reshape into, from what ***IT*** has already reshaped into, (and I am now thinking of the process that we are going through), ***IT*** may choose to use this information to reshape into something else, as one more possibility as a different, new Big Bang, but still within the same moment that ***IT*** exists as omnipresent.

It would help you to see this better if you remember that ***IT*** was only ***ITS*** heated weight that was and is reshaping into the matter that exists within ***ITSELF*** as ***ITS*** constant nothingness.

I have to marvel at the way ***IT*** exists, as both a cold, clear inner and outer shell simultaneously. I am also fascinated by the strange ways that ***ITS*** clear heated weight can change, transform, or reshape into while still existing within ***ITS*** cold, clear nothingness. Another strange thing about the way ***IT*** exists is that these two opposites or extremes exist as ONE (***IT***), in a ratio which is weird. ***ITS*** cold, clear nothingness being 99.99 as ***ITS*** inner and outer body, that has no weight and is a constant; and ***ITS*** heated weight which is less than 1%. At the same time, ***IT*** uses ***ITS*** coldness to pull ***ITS*** heated weight, which can be fragmented and reshaped into infinite arrangements, as suns, or planets that have water, so that you and I, and everything that exists, as what we see as millions and billions and googols of life forms are just ***IT*** as **ONE.**

But returning to ***ITS*** duality: ***IT*** uses ***ITS*** cold to pull ***ITS*** heat, for we should remember that cold does pull heat, and ***IT*** uses ***ITS*** heat as a gravitational pull, to pull ***ITS*** heat back to being less than 1%. You might see ***IT*** better this way: ***IT*** uses ***ITS*** coldness to pull ***ITS*** heated weight, and ***IT*** uses ***ITS*** heated weight to pull ***ITSELF*** back as heat using the gravitational pull that heat has, which includes everything that can exist as ***ITS*** weight, within ***ITS*** nothingness, so that ***IT*** will reunite to become one total dense heated weight again, which will bring us back to when ***IT*** existed as that very dense matter that existed just at the moment of the Big Bang.

Here is one other way to see this: For ***IT*** to bring back ***ITS*** original weight that now exists as fragments that began existing when ***IT*** sent as ***ITS*** weight outwards, in order for ***IT*** to be able to exist as matter, ***IT*** must bring back ***ITS*** fragmented weight into what we now call black holes, which are really concentrated clear hot spots, in order for ***IT*** to bring back ***ITS*** heated weight that now exists as galaxies.

All black holes as matter will seek each other out, so that they can become one again, and when all black holes are one they will have the energy to pull back any and all of the missing, tiny, loose, fragmented matter that is moving away from us. This will complete the cycle that started out as the very dense matter that was sent outwards at the moment of the Big Bang. Right at this instant ***IT*** is being returned to becoming one very dense weight again, as the very dense heated weight that existed at the moment of the Big Bang.

And when ***IT*** has reunited all ***ITS*** heated weight again into one singular point, within ***ITS*** nothingness, ***ITS*** weight will again explode in order to push ***ITS*** weight outwards again, just like at the beginning of the Big Bang. Now as to just how ***IT*** does this igniting to produce an explosion we will have to use the information that now exists, or just continue understanding ***ITS*** ways without knowing how ***IT*** accomplishes things like explosions, for ***IT*** is beyond science, because science came after the mind, and our minds work as ***ITS*** oneness in the electrochemical exchanges that the mind uses to think about how ***IT*** reshaped into us, so that we could see and understand ***IT*** better, even if we as humans are only a human thought, that has to exist within ***ITSELF*** as ***ITSELF***.

And again I must say thank you to ***IT*** for allowing me to exist and be the best I can, physically and mentally, within ***ITS*** existence.

Here is a thought that relates to ***ITS*** hotness and coldness: ***IT*** could be that ***ITS*** coldness is what ***IT*** uses to pull ***ITS*** heated weight outwards into ***ITS*** coldness, while ***ITS*** heat is always looking to reunite back into ***ITS*** *singularity* so as to return to where this weight started out, as one total heated weight, and this happens because ***ITS*** heated weight has the force of pulling back other heated weight, which is what we know as gravity.

⌘~~~~~~~~~~~~~~~~⌘⌘~~~~~~~~~~~~~~~~⌘

Everything that exists has ITS weight and ITS divine nothingness

❀~~~~~~~~~~~~~~❀~~~~~~~~~~~~~~❀~~~~~~~~~~~~~~❀

The beginning or end of the Universe is only ITS weight

When we say or refer to the beginning or ending of this Universe, we are only referring to ***ITS*** weight, or what we call matter. You will understand me better if you remember that this matter that now exists, exists within ***ITS*** nothingness. So using our imagination, if you could see what took place at the moment of the Big Bang, what you would see is that ***IT*** took ***ITS*** less than 1% heated weight, that exists within ***ITSELF,*** and threw this weight outwards, in quantifiable portions, as what we call protons, neutrons, and electrons, so that ***IT*** could reshape into other possibilities as matter. And since ***IT*** threw ***ITS*** less then 1% weight outwards, ***IT*** will then bring back this weight to form chunks of concentrated weight, such as what each black hole has, and ultimately, ***IT*** will then bring back all the different chunks of ***ITS*** weight together using the force that ***IT*** installed as the properties that gravity has as ***ITS*** weight. So continuing on with our imagination, as this can only be done in one's imagination, ***IT*** will bring back the heated weight that ***IT*** has in the form of supergiant stars, and compress this weight into black holes. However, at this stage we cannot be to close to this event, for we too can be consumed in this process of compression. As ***IT*** continues to compress all ***ITS*** fragmented weight into singular points, such as when ***IT*** compresses all the weight that exists as a galaxy, we too will be consumed in one of these galaxies as ***IT*** is being compressed, and eventually all galaxies will become singular hot spots, in preparation for becoming just one hot spot, which will be the same kind of hot spot that existed at the moment of the Big Bang.

Let me add that I too felt the grimness and sadness of these events, but I too have had to accept that ***IT*** runs ***ITS*** show, and my message from ***IT*** is that I should enjoy my existing moment as a gift from ***IT***, in ***ITS*** reshaping of ***ITS*** weight, within ***ITS*** nothingness, and that I should not worry, because we, as life, will mostly be off this planet and far away, when our sun consumes everything on this now existing planet, but for now ***IT*** has let us know that we still have millions, billons, trillions, and maybe even googols of more of Earth rotations to enjoy our existence.

So, at this stage, when the only things left existing within this Universe, (***ITS*** nothingness) are just all the individual black holes, because ***IT*** has returned ***ITS*** weight from matter to just individual black holes, including us, we will not be here to confirm this experientially in any possible way, for we and anything that we have constructed will no longer exist. But using our imagination, what we would see is that ***IT*** took all ***ITS*** fragmented weight that existed as matter, and compressed this weight into black holes, which are really ***ITS*** clear heated weight. Here we are then back to the moment of what we call the Big Bang. Maybe then ***IT*** will let us know how many times ***IT*** has gone through this process, of throwing out ***ITS*** heated weight as fragments and then bringing ***IT*** back to the singular point. After this imaginative excursion, I think you are able to realize why there is no beginning or end to the Universe, for it was ***ITS*** heated weight that fragmented into atoms and then ***IT*** brought this weight back in different stages as black holes, which will all become one very dense heated singular point again, as one huge black hole. This would be the same state as what we have been referring to as the very dense matter that existed at the moment of the Big Bang, for ***IT*** does this as one continuous singular moment as a place, which is really just ***ITS*** inner and outer ***SELF.*** You will also understand why I say one continuous moment if you recall that our human mechanical time system did not exist prior to the Big Bang. You will be reading more about the subject of time in the following section which is all on the history of our human mechanical time system. In the same way, we cannot apply time to the inside of a black hole. In the first place, we cannot see a black hole simply because ***IT*** is a clear form of heated weight, and in the second place, because we or anything that is made of matter cannot exist inside a black hole. For something to exist for us it must have electrons, because it is the electrons that separate each atom's nucleus from the nothingness that exists surrounding ***IT***, and that enable ***IT*** to join to other atoms without the whole collapsing. Black holes are not separated from ***ITS*** nothingness, for they are in direct contact with ***IT***, as if they were brothers to ***ITS*** nothingness.

Any material object we are able to perceive with our sight is also made in ***ITS*** own image, as in being a shell.

Let me explain. When we see or touch something we are seeing and touching the outermost shell of the object followed by ***ITS*** one total nothingness. Then as always, what is inside is ***ITS*** weight.

Another interesting thought: We know that at the moment of the Big Bang ***IT*** threw ***ITS*** weight outwards. So ***IT*** would be most likely that ***ITS*** lesser weighing weight or lighter constituents went father out, and that ***ITS*** heavier weight, stayed closer to where the Big Bang began.

As scientists look more into how ***IT*** exists, they will be able to ascertain whether this is true. But for now, if any one out there has any information as to where within ***ITSELF*** the Big Bang began, please share ***IT*** with us, so we can see where ***IT*** once had all ***ITS*** weight gathered into one point within ***ITSELF***, as ***ITS*** nothingness.

Let us now imagine a different event, something that has not happened yet: If a black hole could explode, would we be seeing the beginning of a new galaxy? The reason I do not say "the beginning of a new Universe" is because there is really no Universe to begin with, for what we call the Universe is really ***ITS*** body or "shell" that now exists as ***ITS*** nothingness, and what we refer to as the matter that exists inside the Universe as a result of the Big Bang, is really ***ITS*** weight, but not as a body. Here again, you, the reader, like me also, will have to readjust the way you are accustomed to understanding things. Let me give you an example: To us, ***IT*** is our body that contains our weight, as weight itself, just like the way we know that an object's weight is what the object is, as its inside, for we never say that the weight of the object is what is outside the object. To us, anything that does have weight, either exists as us or as being outside of us, just as all the weight that exists as matter is out there, outside of us.

But to understand who ***IT*** is, and how ***IT*** exists and appears to our senses at this moment, ***IT*** is better if we look at what those things that we refer to as having weight, as being out there, in a reverse mode, because if you look at ***IT*** as ***IT*** really exists, be it by the name of GOD, or by the name of pure energy, what we will see is ***ITS*** outer "shell" composed of a nothingness, which we can at least visually perceive as the hugeness of what we refer to as the empty Universe. That is "out there" to us.

Let me also add something that I just noticed: It is still a problem for us to say that this nothingness is ***IT***, as GOD, because if GOD created everything, why would GOD create something that is made up of a form of a freezing cold nothingness? So, for us to understand this better we have to stay with what does exist as the nothingness of that pure energy that cannot be created nor destroyed, and my personal experience is that I have just accepted that since no one created ***IT***, this is the way ***IT*** exists.

But let me get back to the way ***IT*** exists now, as ***IT*** appears to our senses, which is that if we say or refer to ***IT*** as being an object, then we could say that ***ITS*** weight is also inside of ***ITSELF*** as the object. ***IT*** is not a problem to call ***IT*** an object, because ***IT*** is something, and this something does exist. The only problem would be that this object does not have an outside that is made of something that needs weight on the outside in order to be seen. Here's an idea that I think will help you understand ***IT:*** When we use the phrase "everything is made in ***ITS*** own image", this is true. Just look at any object that exists and you will see that that object exists within ***ITS*** one total nothingness, as being within ***IT,*** in the same way that ***ITS*** weight exists within ***ITSELF*** as ***ITS*** nothingness.

So as you can see, every thing is made in ***ITS*** own image. Anything that does exist, exists because ***IT*** used a very tiny amount of ***ITS*** weight to form a visual "shell", because anything that you can perceive visually you are able to see only because ***IT*** used ***ITS*** weight as the tiny amount of weight that the protons, neutrons, and electrons in each atom have, which is followed by ***ITS*** one nothingness that keeps the different portions or fragments of ***ITS*** weight that we call elements together within ***ITS*** total nothingness.

Let me give you some more examples. Let us commence with the forming of an atom, which is where ***IT*** starts using ***ITS*** weight to begin forming what we call matter. Now this atom only exists because ***IT*** used a tiny amount of ***ITS*** weight, and placed ***IT*** on ***ITS*** high speed nothingness, to become an electron, which is going to surround a portion of ***ITS*** now fragmented heated weight, within ***ITS*** total nothingness. Please follow me closely now, because this is the way ***IT*** makes things in ***ITS*** own image. As ***IT*** keeps adding more of ***ITS*** total weight as portions, (that we must not forget are all surrounded by ***ITS*** one total nothingness), ***IT*** begins to build up these individual fragmented portions of ***ITS*** total weight, from one portion, as with the hydrogen atom, which only has one proton, one electron, and no neutron, to more clustered portions of ***ITS*** heated weight to form atoms like carbon, calcium, gold and iron, to allow us to visually perceive ***IT*** as ***IT*** exists in the form of ***ITS*** heated weight that likewise exists within ***ITSELF,*** as ***ITS*** "shell". We should never forget that anything that we can see, we do actually see only because ***IT*** used ***ITS*** weight within ***ITSELF***, and that everything we can see is made in ***ITS*** own image because ***IT*** fragmented ***ITS*** less than 1% heated weight into much smaller portions within ***ITS*** total nothingness.

As ***IT*** continued to add portions of ***ITS*** weight, ***IT*** reshaped into human beings, even into you and me, and as ***IT*** continued to add more of ***ITS*** portioned weight outwards, we can now, from outer space, look back at what ***IT*** reshaped ***ITS*** weight into as this planet, which exists within ***ITS*** oneness, as a nothingness where all of these very tiny portions of ***ITS*** weight can exist as something visual for us to see. So remember that as we look in to ***IT*** as ***ITS*** huge body that exists as a nothingness, we can see the other part of ***ITS*** duality in the form of ***ITS*** fragmented weight, that we have given names to, such as: the Moon, Jupiter, Venus, just to mention a few of the things that ***IT*** reshaped some of ***ITS*** less than 1% fragmented weight into, within ***ITSELF*** as ***ITS*** nothingness.

In short, the images that we are seeing are also images of ***ITS*** weight that exists within ***ITS*** nothingness.

******Even we exist as a minute portion of ITS weight******

Who created the Universe?

My answer is no one. First, let us begin with the word who. For some of us the Universe was created by God. However, let us first consider the meaning of the word omnipresent, which indicates that God is in all places at the same time or moment, which implies that God is everywhere in the Universe and in everything in the Universe. Now, if we use the phrase "God created this Universe", God would have had to go somewhere else other than where God exists to find the things that now exist inside of this Universe as matter, and God would have to find the weight that exists as this matter from somewhere else, which means that there would have to be a place somewhere else that can exist outside of this place where God exists as omnipresent.

Next, let us look at this Universe as what is scientifically known as pure energy. As such, we know that this Universe as pure energy cannot be created or destroyed; it simply exists as what we now see as this Universe.

We have to understand that when we ask who created this Universe, it is simply that our minds have grown accustomed to seeing things as being created or destroyed, when in reality nothing, as the pure energy that matter exists as can be created or destroyed. In fact, scientists have empirically confirmed that what is happening is what is known as transmutation, or reshaping, or transforming. So it is only the human mind that will say that God created the Universe, and this is so because the human mind has not readily accepted, that it, as the mind cannot exist if the material that gives rise to the mind did not come from what exists as pure energy, as the weight that the brain and the body, and everything else that exists did not come from this pure energy (God) and as if this energy could be created or destroyed.

Let me mention that I have accepted that I am not a creation, for every atom that makes my existence possible is only possible because ***IT*** exists as what we have come to understand as pure energy and I have no problem accepting that ***IT*** is everything that exists, be it "good" or "bad", be it from white or black, be it hot or cold, for now that my mind has become aware that ***IT*** is everything, I can now just be grateful that ***IT*** allows me to just watch ***IT*** as ***IT*** reshapes into all of these existing possibilities that are happening on this planet as ***ITS*** weight within ***ITS*** nothingness, and to also see how ***IT*** exists as both of these qualities: as what exists as this Universe, as ***ITS*** body, where ***IT*** is just reshaping ***ITS*** weight to ***ITS*** satisfaction, and not to the satisfaction of my mind that depends on my brain which exists as ***ITS*** weight.

Please allow me once again to remind you, the reader, that it is not important who or what may be the maker of this Universe, for one can easily get lost within one's existing moment as life, trying to figure out who or what makes this Universe tick. What is truly important is that who or what makes this Universe possible resides within us, and it is up to our human mind to accept that it too (the mind) depends upon ***ITS*** weight as every atom that the brain is made from, which exists within ***ITS*** divine nothingness. For this reason, I hope that you, the reader, do not get lost in the abstract thought about how things exist outside of you.

For this reason also, I wish that I could find more ways of presenting to you as much as I can about how you can understand ***IT*** better, because I too was once giving more importance to understanding the Universe as being more important than looking to understand ***IT*** directly, as how I and everything that exists is really 100% ***IT*** as omnipresent.

Let me offer you my personal understanding: To me it is good that so much information is available that can help me to learn about ***IT***, be it spiritual information or scientific information, or from people like Maharaji, (www.maharaji.com) or from reading information from other perfect masters that have found ways of understanding ***IT*** better, whether using language or through meditation, because ***IT*** is so beyond our wildest imaginations, so unique, so perfect, so understanding, that ***IT*** allows us to be able to experience ***IT*** both in meditation and as how ***IT*** exists outside of us, which entails how everything is really just ***IT*** reshaping ***ITS*** weight. Now I do not have to attach myself to any one person or thing that may exist outside of me any more, for I know that they too are ***IT,*** as omnipresent.

⌘~~~~~~~~~~~~~~~~~~~~~~~~~~~~~~⌘

******The only one that can hold you to IT, is you.*******

~~~~~~~~~~~~~~~~~~~~~~~~

ITS weight, as matter

Here is another situation where you will have to use your imagination and ***IT*** has to do with ***ITS*** weight, and the way ***IT*** reshaped as matter.

I need for you to imagine the moment when ***IT*** threw ***ITS*** weight outwards within ***ITSELF*** as ***ITS*** nothingness. When this happened ***IT*** had to do something that would keep these individual quantifiable fragments or portions of ***ITS*** weight separated. I use the word separated because we need to remember that one quality or physical law is that ***ITS*** heated weight will always be seeking to return to being just one concentrated singular weight. Now, in order to make this fragmented weight that ***IT*** threw out in the form of an explosion (the Big Bang) stay separated, instead of pulling ***ITSELF*** back together again as one, ***IT*** took just a very tiny amount of this weight, which really belongs to ***IT*** anyway, and combined ***IT*** with ***ITS MAXX-SPEED*** to form the electron, so as to wrap ***ITSELF*** around these fragmented portions, so that they could not join back together as ***ITS*** heated weight.

So imagine this: ***IT*** took ***ITS*** heated weight, fragmented ***IT*** into very tiny portions within ***ITSELF***, and ***IT*** keeps these portions that continue to exist within ***ITSELF*** separated by the electrons, which have both ***ITS*** qualities, so as to allow ***ITSELF*** to exist as the matter that now exists within this Universe, which is really ***ITS*** inside. By using your imagination you are now able to see that what exists as matter are just portions of ***ITS*** heated weight (atoms) that still exist within ***ITS*** huge emptiness. This huge emptiness is analogous to the empty space that exists inside each atom. If ***IT*** were possible for you to look at ***IT*** from a distance this
~~~~~~~~~~~~~~~~~~~~~~~~

is what you would see in this stage of ***ITS*** existence. But you have to imagine that you are seeing ***IT*** as a totality, as what ***IT*** exist as ***ITS*** nothingness, as ***ITS*** shell, body, or housing, so that as you are looking at ***ITS*** cold empty body, (the Universe as omnipresent) you will see that ***IT*** took ***ITS*** once singular heated weight, and fragmented ***IT*** within ***ITSELF***, and keeps this fragmented weight that exists inside ***ITSELF*** as protons and neutrons, and ***IT*** keeps these quantifiable portions of ***ITS*** once singular weight separated, as the electrons.

So that if we could see ***ITS*** weight, (which we cannot, for ***ITS*** weight is also a clear form of being), but let us say we could see the individual protons of ***ITS*** weight that exist within ***ITSELF***, what we would see is ***ITS*** portioned weight that exists with in ***ITSELF***. These individual portions are separated from each other within ***ITS*** one total nothingness. This is why electrons exist: to keep the protons and neutrons apart from other protons and neutrons so as to form separate atoms, so that these portions of ***ITS*** weight can continue to move about within ***ITS*** nothingness, but not be able to reunite due to the force of gravity. These portions do have a tendency to reunite because of gravitational pull, but the electrons do not allow this to happen.

⌘~~~~~~~~~~~~~~~~⌘~~~~~~~~~~~~~~~~⌘

We, as life, have been given a chance to participate in moving ITS weight around **

❀~~~~~~~~~~~~~~~~❀~~~~~~~~~~~~~~~~❀

Understanding pure energy

I have asked friends if they know why pure energy has no beginning or end, and the only reply I've heard is that an energy exists that just does not have a beginning or end, which doesn't answer the question. So I ask ***IT***, "Why?" By using a wall as an example, we know what a beginning and end looks like. This is because the wall is made of matter (weight). I recommend that you, dear reader, again look at the photo of the young and old women. Like that photo, ***IT*** (God, pure energy) operates as a duality. So, when we want to see and understand something, we should look for the other perspective. When I started looking at ***IT*** from a new view, I also asked ***IT*** if I could understand ***IT*** a little better.

I, like anyone else, like to feel that I am standing on my feet when I see and understand things in a different way. I cannot really describe the awakened sense I felt when I received the information on time, matter, and the nothingness that ***IT*** exists as. It was as though someone turned me upside down, a position quite distinct from the one I was accustomed. Anyone can check out what it is like to see something one is familiar with from an upturned position; just stand on your head, and you'll notice that it takes a while to adapt to the totally different perspective. What is important here is that ***IT*** is this extreme view too.

That is why I like the photo of the young and old women; it makes me remember that ***IT*** exists as one but operates as two. Just as the wall exists having a beginning and end because the wall is composed of ***ITS*** weight; ***ITS*** nothingness exists as not having a beginning or an end.

Reactions

In physics there is a law, known as Newton's Third Law of Motion, which states that for every action there is an equal and opposite reaction.

As an example, when a gun is fired, as the bullet goes forward the gun moves back, that is, in the opposite direction, into what is known as the kickback, or when a boat goes forward, the water goes backwards, just to mention a couple. Now, if we apply this to ***IT***, that could mean that since ***ITS*** heat has been pushing outwards ever since the moment of the Big Bang, then ***ITS*** coldness must be pulling inwards.

IT as energy

Here is an idea that arose from my answering a friend who asked me why I refer to matter as energy and not as matter, and ***IT*** is because ***IT*** is more of an energy than matter. The reason for this is that matter is composed of energy, and that ***IT*** is from energy that we get matter, for as I have said elsewhere, matter is ***ITS*** weight occupied by ***ITS*** oneness as the electron.

The way I view ***IT,*** I can understand ***IT*** better if I stay focused on ***IT*** being a dual energy or an energy duality.

Maybe my readers can understand ***IT*** better through Einstein's famous equation: $E=mc^2$, which may be freely rephrased as matter = energy, for they are both interchangeable, because they are both ***IT*** as ***ITSELF***.

And one more reason why I see ***IT*** more as an energy is because when ***IT*** existed as stage #1, ***IT*** existed as dense weight and not as the matter that came from this dense weight that later became atoms, as the matter that now exists.

Robots

Robots exist as a result of ***IT*** reshaping ***ITS*** heated weight into matter, as atoms, and atoms exist only because ***IT*** reshapes as pure energy. First ***IT*** reshaped into us so that we, with our eyes and hands could build robots in our image, having eyes and hands, too, and being able to reshape what exists outside of itself as a robot, to the now existing point where robots are producing more robots, for some of us have seen our jobs replaced by robots.

In this I first acknowledge my gratitude to ***IT*** that I am still here in this moment of ***ITS*** existence, where I can at least see what ***IT*** continues to reshape into, for I, like you, cannot stop what ***IT*** will do; we can only watch what ***IT*** is going to reshape into as ***IT*** uses what ***IT*** now exists as. When we, as ***IT***, leave this planet in order to continue surviving as living beings, we may be accompanied by robots that will do the things that we cannot or should not do as humans in outer space.

Visual mass

As for mass, when we look at some thing we need to know that something exists in order to have a mass, or extension in space, yet one reason why I feel it is better to talk about ***ITS*** weight, which scientists refer to as ***ITS*** mass, such as the mass inside an atom, is because only 1% of ***IT*** exists as mass. The other 99% exists as a nothingness which cannot be spoken about as having mass, yet does exist as having distance and exist as an energy that exist as a freezing cold temperature.

⌘~~~~~~~~~~~~~~~~~⌘⌘~~~~~~~~~~~~~~~~~⌘

*****As for energy and mass being interchangeable, IT is true, for we are in both cases referring only to ITS heated weight.****

How dense are you in terms of ITS weight?

As we reshape, questions about ***IT*** emerge: Are we permitted to know how wide ***IT*** is as this Universe? And how heavy ***IT*** is as weight? I would not dare to say that IT is fat and needs to diet, but we now know that ***IT*** carries ***ITS*** weight in an area that is empty and transparent and does this in order to move at high speed, to spin, and reshape. ***IT***, as less than 1%, weighs millions, billions, trillions, or more tons. And ***IT*** uses color to see that which ***IT*** reshapes into, since what ***IT*** reshapes into is empty. But what if we could remove emptiness?

Can we know how dense we really are?

If you are interested in knowing how dense you are, that is, how much smaller you would be if you could remove all the emptiness that is you, there are various ways an experiment could be done. Here are some do-it-yourself ways:

The most scientific way would be to find a small square water tank that holds, say, 100 gallons when filled to the top. A square tub will enable you to calculate the volume of water in cubic inches, since one cubic inch is equal to .004 gallons. We also know that one-gallon weighs 8.3356 pounds. We need these measurements of volume and mass for the experiment, for density equals mass divided by volume. Here is what you can do:

Fill the square tank and get into it. You will displace the water in the tank with your body's occupied volume, forcing the water in the tank to overflow. Then, get out of the water tank, leaving behind as much water as possible. You will now see what your body's volume is approximately by the water that is missing--you just have to calculate the amount of missing water. This amount will be your volume in relationship to the amount of space that your human body occupied as water, which is what you need.

You can find your weight by weighing yourself on a scale. After determining your body's mass and volume, you then reduce the measurement of volume by 95%, which corresponds to the 95% empty space that exists in every atom.

If you do not have a square tank, you can do the experiment in a bathtub; sorry, this experiment will not work in a shower. Get into the tub and fill it with water until it covers your whole body, then mark the water level inside the tub where your submerged body put it. Then get out of the tub leaving behind as much water that is on your body and your hair as possible. You will see where the water level in the tub went down, yet to where it lowered is not important.

What is important is that you find a one-gallon jar. Use the jar to replace the water that is missing, counting the gallons that it takes to fill the tub to the line where you made the mark that covered your body when you where submerged at the beginning of this experiment. You will then know how many gallons it took to reach this line. To determine how dense you are, convert the gallons into cubic inches (one gallon equals 231 cubic inches). After you have multiplied the number of gallons by 231, now remove 95% of the cubes, so that the remaining 5% of the cubes are approximately what you exist as in dense weight.

The other way to conduct the experiment is on paper, which was my approach since I do not have a water tank or tub, only a shower, and I still wanted some idea of what I would be if I removed the emptiness that exists in my human body, and compressed it to how ***IT*** is as dense matter. For the sake of simplifying the experiment, I will use the hypothetical weight of a 100-pound person.

Since one gallon is equal to 8.3356 pounds, by dividing 100 by 8.3356 we know that 11.9967 gallons is equal to 100 pounds, but let's round up to 12 gallons being equal to 100 pounds to stay with whole numbers. Our math will not be the most accurate, but the results will be close enough, as you will see.

The next step is to determine the amount of mass that a 100-pound person has by converting gallons to USA cubic inches. Since one gallon is equal to 231 cubic inches, then 12 gallons is around 2772. Now all we have to do is mathematically remove 95% of this 2772, which was just empty space. This will leave us the 5% that exists as the human body's weight. Yet here again we must rethink our accustomed way of understanding matter--that matter is what we see, because matter in reality is just ***IT*** as weight. However, we started this experiment with the illusion of seeing things as matter, so let us for now continue this way.

We compress ourselves into the 5% of matter that weighs the 100 pounds and takes up an area (minus empty space) of 138.6 cubic inches, which would resemble a stack of 138.6 cubes that would be 138 inches tall, and one inch wide and one inch long, or since that would be nearly double the average person's height (11.5 feet), we can imagine two stacks of cubes 69.3 inches high (about 5 feet 8 inches), that are two inches wide and one inch long. We can also picture this new density as one big cube having equal dimensions of 5.2 inches per dimension (5.173'=138.6 cubes) that still weighs 100 pounds. The next time someone suggests that you are too big in size in relationship to your weight, you can reply that as a person weighing 200 pounds you are only the size in actual density of a cube measuring 6.5 inches in length, height, and width.

If this experiment were done in a more controlled environment and with more specificity, we would know precisely the size as pure energy that exists as weight when a human body is compressed. Using the human body to understand the density of matter helps me to understand ***IT*** better.

IT is not as Dense

I am no mathematical expert, so I will pass this on to the pros. We were able to reduce the density of a hundred pound human body to approximately 5.2 cubic inches, but how does this compare to the density of ***IT,*** as say, one cubic inch of the original dense matter of the Universe, with its tons per square inch? The comparison may be pointless, infinitely smaller than the tiniest dot you could make. But how magnified does it have to be so that we can see it? Knowing the answer will still not permit us to change the total weight of ***IT.*** This calculation only gives us a better idea as to how much space ***IT*** occupies as dense heat, which could then help us visualize ***ITS*** total size as ***ITS*** less than 1% as how ***IT*** exist as heated weight. We might also then understand how ***IT*** adds heated weight to form matter.

ITS weight

We see some of the things that ***IT*** does with its weight; metals that are very hard, heavy and difficult to move, oxygen, and other gasses that float. Yet, all that exists came from this original mass, that scientists have estimated weighed as much as tons per square inch, and on this note, I disagree on the basis of the experiment above. And let me also say that to me it makes no difference which density is right, for ***IT*** will always be independent of what I think. So, as strange as it may seem, ***IT*** is the weight and nothingness that permits me to place this information where you can read it.

We are accustomed to seeing everything that exists within this Universe as matter, when in reality it is ***ITS*** weight that we are seeing as matter. Before the Big Bang, matter did not exist, but ***ITS*** weight did exist, for it was ***ITS*** weight that, after the Big Bang, became what we now call the Periodic Table of Elements, or matter.

How much of ITS heated weight do we gain?

I would now like to share my thoughts relating to the amount of weight that we gain as ***ITS*** weight, and we can see this if we start from the very moment we are conceived. In just 9 months inside of our mother's belly, we can gain an average of 10 lbs (160 ounces in 270 Earth rotations, in terms of days). This means that we are gaining a little less than 2 ounces as ***ITS*** heated weight for each 24 hour period during these 9 months. After we are born, let us say you have become 20 years old (or to say it my way, you have existed for 7305 days in terms of the rotations that our planet makes around the Sun), and you weigh 150-lbs. This would mean that you gained another 140lbs of ITS heated weight in these 20 years. In short, this means that you gained an extra one third of an ounce, after you used what heated weight you needed as energy for transferring ***ITS*** weight while engaging in your daily activities, such as work or play. So we could say that we gain approximately one third of an ounce for every Earth rotation as stored energy.

Knowing this, I understand why we don't notice as most things gain ***ITS*** weight, for it is hard to see anything gaining less than 2 ounces of body weight for every time this planet makes one rotation

**** *ITS nothingness is never inside of ITS weight* ****

However, after we have matured, gaining just one quarter of an ounce daily will get us into trouble, because this small amount will mean that after the 150 lbs that we were already weighing, we will gain an extra 5 lbs yearly, or 55 lbs in ten years or 110 lbs in 20 years, and this extra weight will not be visible in our daily existence. For this reason, after one reaches “a certain age” one should become aware that one has to reduce one’s intake of ***ITS*** heated weight, or work more to burn this heated weight off in terms of calories. Blessed are those people that ***IT*** reshaped into as ***ITS*** possibilities of existing, that are always slim!

ITS weight is always inside of ITS nothingness, not vice-versa.

IT is not normal

For something to be normal, there needs to be another thing to compare it to, and nothing compares to ***IT***. In addition, in order for us to understand something, we first must accept that something as something that exists. So here again, the human mind must work to understand ***IT***, for ***IT*** is mostly nothing. A great teacher called Maharaji, who teaches meditation techniques for reaching and coming into contact with the inner-self, asks his students who are learning the techniques, "What do you see or feel?" And when students answer "nothing," Maharaji asked them, "Are you sure?" and when they answer "Yes; I see nothing," Maharaji answers, "GOOD." Students of Maharaji hear why the human mind cannot understand this knowledge, and why, when we have tried to connect to the nothingness of our inner self, our mind keeps disturbing us: Our mind knows it is going where it cannot understand; our minds resist connecting to our nothingness because our minds cannot control nothingness. ***IT*** is not normal. And I am sorry that I personally cannot offer you, the reader, more on this nothingness that exists within us, for there are few words for the human mind to use to describe ***ITS*** nothingness beyond Pure Energy, God, and omnipresent. We can however experience nothingness as an existing moment. And when one does connect with this nothingness for a moment, ***ITS*** effects last even after we disconnect and return to our "normal" way of existing. In fact, it becomes our choice to return to normality. We also have the choice to surrender to ***IT***, which is what I have done. I discuss this in the section of the book called "The Uncertainty Road." Here I explain how I asked ***IT*** to show me what I am supposed to do, show me the road I am to walk.

⌘~~~~~~~~~~~~~~~~~~~~~⌘~~~~~~~~~~~~~~~~~~~~~⌘

The Buddha found a stage of meditation that exists as a nothingness.

❁~~~~~~~~~~~~~~~~~~~~~~~❁~~~~~~~~~~~~~~~~~~~~~~~❁

I will never change the road that ***IT*** has permitted me to travel on. Now I say to ***IT***, "Please never kick me back to that road considered the 'normal' way of existing."

IT as an eye

Here is another idea that might help you in seeing ***IT*** better. This comes from my questioning why we refer to the third "eye" that exists within us. One reason for this is that we already have two eyes to see outwards, but we also have an eye to see inwards, which is only for seeing ***IT.*** To this I should add that this third eye is not really an eye. Calling ***IT*** an eye is just an expression to describe something that exists where we can look into ***IT***, and I say look into ***IT*** because again, we cannot look or exist outside of ***ITS*** existence. The best analogy I can think of to describe this is that of a body cell, which is basically light and thin on the outside, something like ***ITS*** nothingness that exists as ***ITS*** outside, and like the cell where the cell's weight is in the inside, which is the same way ***IT*** exists, by having ***ITS*** weight in the inside of ***ITSELF.***

Why we cannot go through IT as ITS weight

To begin, let's review some information we already know. I would like to use a hydrogen atom as an example. The hydrogen atom is the simplest one, being made up of only one proton and one electron. The proton composes the nucleus, which is separated by 95% empty space, from the atom's sole orbiting electron.

Now, I have tried to understand ***IT*** and ***ITS*** ways of doing things, and we already know that ***IT*** has a way of making ***ITS*** weight behave as a repellent to the ways that ***ITS*** weight exists as. To clarify, all protons are made of ***ITS*** weight and we know that one proton will repel or push away another proton because they are of the same polarity. This is one of the ways that ***IT*** functions.

Electrons also repel each other, for the same reason. For instance, when one object gets close to another their electrons may also be repelling each other.

Let me now share with you what I see as a way of ***IT*** doing something: We know that all matter is made up of atoms and all atoms have at least one proton in their nucleus, [to me, protons are fractions of ***ITS*** once singular weight], followed by the empty space that exists inside every atom, [to me all the emptiness inside every atom is the same one total empty nothingness that ***IT*** exists as, as ***ITS*** shell body] and after this emptiness that exists inside every atom, comes the outer part of the atom, where the electrons are. The function of the electrons is to hold in what exists inside the atom as distinct portions of ***ITS*** fragmented weight. But electrons [which are made up of both ***ITS*** speed and a very tiny amount of ***ITS*** weight] also have another function, which is to keep out other portions of ***ITS*** weight.

And since these electrons do exist as something having exterior polarity, this might be a way that ***IT*** uses to stop or push away other matter that may try to penetrate the atom's exterior. If this were not so, the interior of the atom would be thrown into disorder, atomic and molecular organization would break down and matter as we know it could not exist.

So, for example, when we put our feet on the ground, the outermost layer of both your feet and the ground being made up of electrons, they may be pushing each other away, so that the atoms of our feet and those of the elements making up the ground do not mingle. I must add that what I feel is really happening is that ***ITS*** weight in the form of electrons does not let us penetrate material objects made up of ***ITS*** fragmented weight that ***IT*** has reshaped into as atoms.

As to why this is so, ***IT*** may be building up a shield, something like a wall, as the electrons, and also, there may be pressure inside the atom pushing outwards from inside, so as to act as a wall, or shield preventing penetration and therefore a breaking down of the atom's internal order.

Now my observation is that when I try to penetrate an object's outer layer, where the atoms' electrons are, to reach the 95% emptiness that exists inside, I experience resistance. This is so because of the way atoms are built, having ***ITS*** weight as electrons on the outside and then ***ITS*** weight as protons and neutrons inside the atom's nucleus. These two portions of ***ITS*** fragmented weight are separated by a 95% emptiness.

Now let me share with you my feelings concerning what may be happening relating to ***ITS*** nothingness. We know that ***IT*** exists as one huge, empty nothingness, where ***IT*** reshapes ***ITS*** weight inside of ***ITSELF***, where ***IT*** can take ***ITS*** weight and reshape ***IT*** into what we now see as matter. But in order to continue my explanation I will have to repeat some things we have already discussed so that you may be able to see things from a different point of view.

When I see an object, I know that this object is ***IT***, as ***IT*** used ***ITS*** weight to exist as matter. This matter exists as having weight and all these portions of ***ITS*** fragmented weight are prevented from re-uniting with other fragments of ***ITS*** weight with a given distance. This distance is the distance that exists inside each atom between the nucleus and the orbiting electrons. We refer to this distance as the 95% empty space that exists inside each atom.

So, knowing this, whenever I start trying to push my way into any material object, I see that I may be able to push the atoms to the side, but I cannot go through them. One good example of this is the oxygen in the air. Oxygen in Earth's atmosphere exists as a gas composed of molecules made up of two atoms. When I move from one place to another on this planet, my body pushes the oxygen molecules to the side, but my body cannot go through them. If oxygen in the atmosphere existed as a solid instead of a gas, my body would bump into it the same way it bumps into a cement wall. As ***ITS*** weight, it is easier to push them aside because they are in gaseous form, since the molecules are farther apart from each other. But when I try to push something heavier or denser, such as a liquid or a solid, where the molecules are closer together, the harder it is to push the atoms aside. And while trying to understand this situation I came to the conclusion that I can only push or move ***ITS*** weight as the object itself.

I went so far as to watch what I, as a human being, could do in trying to penetrate ***ITS*** weight: I tried to push my finger through the table, knowing that my finger, and the table both exist as having an outer layer of electrons and that it is the weight that exists inside of these electrons in my finger as my skin, finger muscles, and finger bones, while the same thing is true about the table. Of course, I found the table would not allow my finger to pass through because ***ITS*** weight, as the atoms the table is made of, also have an outer layer of electrons.

So my conclusion is that so that matter could exist, ***IT*** first had to form atoms, and this ***IT*** accomplished by placing a tiny amount of ***ITS*** weight as the electron, to stop what ***IT*** exists as in the weight inside the atom (protons, neutrons) from getting out, and also, to prevent any other fragments of ***ITS*** weight from getting in and changing the way ***ITS*** weight may exist as, for it is in the way ***ITS*** weight exist as, that matter can exist in the infinite ways that ***IT*** can change through the reshaping or transforming of ***ITSELF***.

I also realized that the more weight ***IT*** places as an object, as ***ITSELF***, the harder ***IT*** will resist outside penetration of any other portions of ***ITS*** fragmented weight.

Let me try and give you, the reader, a different view: It is not ***ITS*** nothingness that is resisting, for ***ITS*** nothingness, as nothingness, does not exist as something that has resistance, for it is ***ITS*** weight that exists as something, so that it is ***ITS*** weight that is giving what ***IT*** exists as matter the quality of resistance.

We know that a hydrogen atom (element #1 which has only 1 proton, 1 electron and no neutrons) is the one into which ***IT*** has placed the smallest amount of ***ITS*** weight to form matter, and that gold, for example, has more weight than hydrogen per atom, in both the nucleus and the electrons, but the empty nothingness that exists inside all atoms is still there. ***IT*** places more of ***ITS*** tiny weight as electrons in a gold atom so as to keep in this weight that gold exists as, compared to hydrogen, and the weight that exists as the electrons will also keep out other fragments of ***ITS*** weight that may exist outside this gold atom (element # 79, which has 79 protons, 79 electrons, and 118 neutrons), for we know that if more weight is added to this gold atom it will become mercury (element #80, which has 80 protons, 80 electrons, and 120 neutrons).

This shows us that as ***ITS*** weight changes, the object changes, and so does ***ITS*** outer weight change as electrons, which will not be easier in allowing any other portions of ***ITS*** weight to come in.

The only way to penetrate the way ***IT*** exists as atoms, composed of protons and electrons, is by using one of those huge atom smashers that scientists call Particle Accelerators.

The electrons that form the outer layer of atoms are in constant motion around the atom's nucleus. Electrons in motion create a magnetic field and it is this force that operates when for example you put your feet on the ground. The magnetic field created by the electrons in the atoms that make up your feet tissues are pushing away from the magnetic field created by the electrons in the atoms of the elements that make up the ground you are standing on. The same thing happens when you lean on your hand against a wall.

However, my physical sensation (and I repeat, it's just a feeling, for this is all I have to go on at times, as I experience my existence on this planet), is that when I put my feet on the ground it is not pushing away. But my feet are just not going through the ground because of the weight that the ground exists as; that is, the ground itself as the outer layer of ***ITS*** weight as the electrons that are holding in ***ITS*** weight as atoms.

I have tried approaching this situation on an equal basis, in which I could find something that would be equal to my feet or hands, as to what my feet would be touching, or my hands, coming in contact with, such as a wall, as ***ITS*** weight, so that these magnetic fields would be equal, to see if I could physically detect these magnetic fields repelling each other, as sensations that I could feel in my fingers or feet, but so far, no such luck. I just keep feeling that it is ***ITS*** weight that keeps stopping me from going through the ground, following the force of gravity all the way to the Earth's core, or from my hands going through the wall into the adjacent room, and it is because of the weight of my finger and the wall that keep stopping each other. However, as far as physical sensations go, I have never felt a magnetic field repelling me. Who knows? Maybe it's just me. So, if anyone reading this finds a way of feeling any sensations produced by this magnetic field, let the rest of us know how to experience this feeling, without hitting your head on a wall and getting a lump or getting all bruised up.

I also feel that if I could penetrate this first layer of ***ITS*** weight (where the electrons are), I should be able to continue penetrating the way ***IT*** exists as ***ITS*** nothingness, because as ***ITS*** nothingness ***IT*** does not have resistance. This is something we already know about concerning the way ***IT*** exists as what we call outer space, which is where ***IT*** exists as a nothingness, which we know offers no resistance. It has even been commented that if we could throw an object into outer space (***ITS*** nothingness) and if this object did not hit any other object (as ***ITS*** weight) the object thrown would at some moment would return to its starting point again because there would be no resistance as this object traveled through ***ITS*** nothingness.

Nevertheless, the weight ***IT*** put into electrons is so strong that we cannot go **through** this barrier into which ***IT*** reshaped in order for ***ITS*** weight to be able to exist as atoms. Furthermore, we have to be grateful to ***IT*** for this way of being, otherwise, I could not write about these findings and you could not read about ***ITS*** weight that exists within ***ITS*** nothingness.

It seems that the only way we can safely experience ***ITS*** nothingness is through meditation, for we cannot step directly into ***ITS*** nothingness, as outer space, without special clothing, and as humans, we can, depending on the climate, connect with ***ITS*** nothingness as meditation without wearing any clothing. So, it looks like there is no way to just step into ***ITS*** nothingness.

To continue, I know that there is information on this existing magnetic field that states for example that however close our bodies come to another physical object, there is not any actual contact. What happens is that one's body gets close enough for there to be a transfer of energy back and forth. Scientists in fact argue that one's body and the object in question actually repel each other and this is what prevents us from passing through physical objects. This may be so, but when I try to walk through a wall, I do not get a repelling feeling. However, if I try hard enough I do get bruised up!

Let me mention that yes, there are some things that do go through us however. A good example is X-rays, but this is not a healthy way to experience ***ITS*** nothingness...

However, it is true that magnetic fields can prevent things from coming together. I still remember this experiment from science class in elementary school. The teacher made us take two bar magnets and try to bring like poles together (that is, north to north or south to south). As hard as we tried the closer they came to each other the stronger they repelled each other. Wondering whether my fingers or feet have a magnetic field, I looked around my house, but I did not find any objects that I could practice with. If my fingers do have a magnetic field to them, be it positive or negative, I have not yet found the opposing force to my fingers where I can feel this magnetic repulsion. This suggests that the magnetic field that surrounds our bodies must be weak, or that our senses are not equipped to detect it.

Another thing that came to mind concerning the existence of this type of repelling force has to do with when I find a person that is what we refer to as "being negative". When this happens I feel that I should not try to get close to this person, for it is the one time that I can feel this negative force that the person exists as, and it is best that I do not even try to get near this person's negative repulsion field.

Finally, if you, the reader, have any information to contradict or support what I have written above, please send it to my e-mail where it can be shared, (if you so desire it) with the rest of us, because I can still learn from seeing things from a different angle, for I wish that there was a way of bringing together as much of the information that exists in the different minds that exist out there as this moment of ***ITS*** existence so that we could understand ***IT*** better just as ***IT*** existed from before the Big Bang until the way ***IT*** exists presently.

Imagination

Here is one more way to open your mind to how ***IT*** may exist: Imagine that you are God, and that you are everywhere at the same time, and that everything that exists is you, as one as if you were able to be in all places at the same time (moment). In this exercise of the imagination, it will help if you remember that you also exist as one total heated weight which is less than 1% , that exists within a one huge, cold, clear nothingness, as your body as a place where you keep your heated weight with inside of it (here I am likening your body to this empty Universe), that can exist as just one total weight, (like at the moment of the Big Bang) that can also be fragmented in to very tiny portions (such as protons and electrons).

It will help if you remember that you are God as a portion of ***ITS*** heated weight that exists within a portion of ***ITS*** total nothingness. Now, as to what would you see and comprehend, if you find that you would like to share it with the rest of us, please send it, so we can use it in understanding ***IT*** better.

Why IT exists as Humans

Here is an interesting thought that I will try to introduce using our human way of existence. From the moment we are born we each need the presence of someone else besides our self, beginning with our mother. As we grow, we need to know that other humans exist, and eventually, most of us have a need to share our lives with a partner. In short, we exist because others like us also exist.

It is self-evident then that ***IT*** is strong enough or complete enough in ***ITSELF*** to exist without a partner or even without the existence of something other then ***ITSELF***, for there is no other such as ***ITSELF.***

We, on the other hand, by observation and experience know that we do need to have communication with others. But by the same token, observing the way ***IT*** exists, one must first realize that there is only one of ***IT***, and that everything that has happened in the so-called "past" or will happen in the so-called "future" is happening to ***ITSELF***, as one moment of ***ITS*** existence. Furthermore, it is within ***ITS*** constant nothingness that ***ITS*** weight exists, and ***IT*** uses this weight to form things that are separate entities. For example, we exist because we are made from fragments of ***ITS*** weight, and we communicate with other humans or interact with objects that also exist as fragments of ***ITS*** weight. It is also obvious that this is what has been taking place from the moment of the Big Bang up until this moment of our existence, for science has confirmed the process.

So returning to understanding ***IT*** a little more in terms of how ***IT*** exists as just ONE, we can say that ***IT*** is using ***ITS*** weight that exists inside of ***ITSELF*** in a graceful manner to further ***ITS*** own agenda. You will have noticed that I have used the word "graceful" to describe ***ITS*** reshaping process, because having thought about the way ***IT*** first transformed ***ITS*** pure energy into atoms in order to reshape into everything that exists inside of ***ITSELF***, eventually providing everything that would be necessary and setting up the platform for human life to appear seems to me a very graceful thing. And now that we are here, we can confirm that ***IT*** exists. I am thinking of the countless times that people, who are made up of fragments of ***ITS*** weight, in despair or suffering the ravages of war have called out to ***IT*** for help and have received what they needed. I am also thinking about those other fragments of ***ITS*** weight called priests and monks who have dedicated their entire lives to coming to know ***IT*** more intimately through prayer and meditation.

There is some food for thought here: It seems that ***IT*** has programmed humans to communicate with ***IT***, because other fragments of ***ITS*** weight, such as animals and inanimate objects do not seem to have this ability or capacity to confirm that ***IT*** does exist. For this reason, and for many others, I must personally say that I am grateful to ***IT*** for allowing me to understand what little I have learned about how ***IT*** exists in terms of ***ITS*** weight that exists moving around within ***ITS*** freezing cold, clear nothingness, which is what we know as this empty Universe. To me it is something truly wonderful that ***IT*** used ***ITS*** weight to form this planet, with its seas and dry land, where I could live and participate in observing the way ***IT*** reshapes ***ITS*** weight, so that I could find my way back to confirming ***ITS*** existence, and telling others about what I have learned by writing this book. So again I say to ***IT***, thank ***YOU*** for allowing me to exist as this peaceful moment that ***YOU*** have permitted to exist in as a portion of ***YOUR*** weight.

⌘~~~~~~~~~~~~~~~~⌘ ⌘~~~~~~~~~~~~~~~~⌘

*** ***IT has no mate. IT remains eternally single because IT is complete in ITSELF.*** ***

❁~~~~~~❁

The ways ITS weight exists as water molecules

Before I start this section let me begin by saying that you will be reading mostly about the different ways that water exists as, but more important, is that you try to see, how ***IT*** has reshaped ***ITSELF*** as ***ITS*** weight as water, as one of ***ITS*** existing possibilities, so that ***IT*** could do many of the things that you will be reading about as ***IT*** as water. We all know that water is the most common substance in which hydrogen bonding occurs, and that water molecules are composed of one oxygen atom and two hydrogen atoms.

We are also aware that all life forms as we know them cannot exist without the way ***IT*** has reshaped ***ITSELF*** as water, and water is our most important natural resource that exists as ***ITS*** weight.

IT as a liquid. I would like to begin by saying that we cannot survive for more than a few days without ***IT*** as ***ITS*** presence as water, and we cannot survive for more than a few weeks without ***IT*** as ***IT*** reshaped ***ITS*** weight as the food we have to consume. Yet as water, ***IT*** gave us less than one percent of all the water found on our planet, as water that is readily available for our nutritional use. I am talking about fresh water because we cannot consume salt water from the oceans. This is the water ***IT*** provides us in wells, rivers, and fruits. These sources supply us the liquid necessary to sustain our body's way of existing.

Let me explain this better: One reason why we exist is because of the water that ***IT*** reshaped into as ***ITS*** weight, so as to enable the existence of what we call a living cell. The first living cells came to be when ***IT*** reshaped ***ITS*** weight as water into oceans, which is most likely where ***IT*** began to reshape ***ITS*** weight into the 92 different naturally occurring elements that have been classified in what we know as the Periodic Table.

We have tons of information on how ***IT*** exists as ***ITS*** weight (atoms-matter) in the oceans and as the many ways that ***IT*** took ***ITS*** weight to become the many living things that exist having living cells, which are mostly made up of water, forming their bodies which could not exist if they hadn't been made by ***ITS*** reshaping into cells.

From my point of view, ***IT*** is everything that exists that is made of matter. Matter is really just ***ITS*** weight, which is what ***IT*** uses to give anything that can exist a shape or form, so that something physical can exist.

Now, let me take you from this point of something being physically alive and use our human physical bodies as an example. They could just not exist, if ***IT*** had not started reshaping ***ITSELF*** as ***ITS*** weight, into the life that ***IT*** exists as, as being ***ITSELF,*** as one, that we know that exists inside the oceans, that we call trillions of living things, which are really just ***ITSELF,*** as one.

I feel that the reason why ***IT*** started ***ITS*** reshaping of ***ITS*** weight as ***ITSELF*** as life in this planet's oceans is because by the time ***IT*** reshaped ***ITS*** weight as this planet, ***IT*** had already reshaped ***ITS*** weight as the Sun, which is one of the ways that ***IT*** sends out ***ITS*** fragmented weight as heat, and this fact is of great importance in understanding ***IT*** as what ***IT*** reshaped into as living forms.

So ***IT*** first had to reshape ***ITSELF*** as ***ITS*** weight as water and place ***ITSELF*** into the 3 different states that water exists, and ***IT*** then had to carve a deep hole on ***ITSELF*** as this planet, probably reshaping through volcanic activity, using lava to reshape ***ITS*** weight into land and then maybe send a portion of ***ITSELF*** as a giant meteor or an asteroid to carve a deep hole on this planet so that ***IT*** could capture ***ITS*** weight as rain into a given area, as the cavities that now hold all of ***ITS*** water as oceans.

After accomplishing this, as ***ITSELF*** as ***ITS*** weight that exists in the oceans, ***IT*** could begin to place more of ***ITS*** fragmented weight inside ***ITSELF***, as one way of existing as the life that ***IT*** exists as. This it achieved by becoming the first living cell, which was really just a fragment of ***ITS*** weight reshaped, having the capacity to make copies of itself, which later developed into all the different life forms on our planet. All these life forms are just ***ITS*** one life force, that ***IT*** exists as.

So we can see that in fragmenting ***ITS*** weight as a cell, ***IT*** then uses ***ITSELF*** to reshape into what we see as ***ITS*** weight in the form of tissues, skin, bones, and muscles, which will then hold in other portions of ***ITS*** fragmented weight in the form of organs, that include the liver for cleansing, the heart for pumping the blood, and the lungs for transferring ***ITS*** weight as oxygen into energy, so that ***IT*** could exist as the most powerful water based organ that has ever existed: the human brain, that you and I exist as.

Our survey clearly shows that ***IT*** started all of this reshaping into cells from beneath the oceans, where the first cell could be protected from ***ITSELF*** as the direct solar heat that ***IT*** also exists as, and it was necessary that ***IT*** start this reshaping into life underwater because ***IT*** is always sending out ***ITS*** fragmented weight as solar energy. Therefore, ***IT*** protected ***ITSELF*** from this heat by keeping the first cell where ***ITS*** solar heat would not dry out what ***IT*** was reshaping into as a water based cell, for we know that water is the substance that can absorb the most heat.

Now we know that ***IT*** reshaped ***ITSELF*** as ***ITS*** weight into what looks like trillions of life forms, which are really just ***IT*** as the way ***IT*** exists as just being one, as the way ***IT*** exists as what we understand as life.

I also know that if you are like me, it is hard to see everything that exists as life as just being one, but you will see this better if you remember this:

1- That ***IT*** (GOD) created everything, and more important, GOD is in all places at the same moment, as in being in every atom that makes every living cell possible.

2- That science bears witness that everything that exists is just this one pure energy, which cannot be created or destroyed, for this pure energy is just reshaping, transmuting, into everything that exists, including what we call life, be it below the oceans or above, on land, as everything that is alive, or whatever else may exist as life inside of ***ITSELF*** in what we call outer space (aliens), because this is all happening to ***ITSELF*** as the weight that ***IT*** has, inside of ***ITSELF*** as omnipresent.

There are many ways to look at things and so that you will have a different view I ask you, the reader, to try to mentally detach yourself momentarily from the human race and just see everything that exists, first and most importantly as ***IT,*** for ***IT*** is everything that exists, including what we know existed before any life existed, as in the moment of the Big Bang, and all the things that now exist as being alive, or what ***IT*** may decide to exist as for we should remember that it was ***IT*** that allowed us to exist and allows us to keep existing, using ***ITS*** weight to give us a shape (body), within ***ITS*** nothingness.

So let us return to water, but this time I would like you, the reader, to see water as ***IT***, and what ***IT*** has done with ***ITSELF*** as ***ITS*** weight as water. We have already seen that in order to begin existing as matter ***IT*** began reshaping into elements, the first and most simple of which is hydrogen. After reshaping into the next 7 elements ***IT*** reshaped into element number 8, which is oxygen. Now the stage was set for ***IT*** to attach ***ITSELF*** as hydrogen to ***ITSELF*** as oxygen, in order to reshape into water (H_2O). This was the first step in reshaping into what we call life. Then, in and with water, ***IT*** also took ***ITSELF*** as the other elements that ***IT*** exists as, to reshape ***ITSELF*** into a living cell, so that ***IT*** could continue to add more of ***ITS*** weight to begin to form life with tissues, so that ***IT*** could continue to add more of ***ITS*** weight to form larger and larger organisms. For the next step ***IT*** used ***ITS*** weight as calcium to form a skeleton so that ***IT*** could do other things that required calcium as the bones that are needed for ***IT*** to have a different type of body, that in some cases, like ours, can stand upright, having legs that can move on land, and hands that can build, and a brain that can assimilate information. At the present stage of our existence we have accumulated enough information, since when we were living in caves up until now, that we are able to understand that we live on a planet in what we call outer space. This journey was also accompanied by technological inventions such as the huge machinery that existed in the Industrial Revolution all the way to the new industrial nano-machinery that is being built at the microscopic level. The most important development in this journey is that humans have minds that can confirm that ***IT*** does exist, for I have asked myself, if the human mind did not exist, who would confirm that ***IT*** ever existed?

Now that I have mentioned the mind, I would like you to think about this: Since ***IT*** is everything that exists, we should remember that we are ***ITS*** creation (reshaping) as the way ***IT*** rearranged ***ITS*** weight, into all the elements that we are made up of, and ***IT*** performed this feat in order for us to see, read, and think. We must always remember that every particle of matter that we are or use, is still ***IT*** as ***ITS*** weight, so that for me to even exist, I have to say thank you to ***IT***, for allowing me to exist as ***ITS*** weight, as the body that I have, and as every atom that my mind is made up of, and as the energy that I have to use as ***IT*** to even be able to think, for thinking is just a thought and this human thought is also ***IT***, as the energy that this thought exists as.

Now that you understand that everything that exists is ***IT***, and that ***IT*** or GOD is pure energy, that cannot be created or destroyed, consider this saying: "The big fish eats the little fish". Since all the fish are really just ***ITSELF*** as what ***IT*** reshaped into, when a big fish eats up smaller fish what is really happening is just an exchange of pure energy. This pure energy that is ***IT*** is just transforming ***ITSELF*** as one fish eats up the other fish, for they are both ***IT***, as GOD and as pure energy.

So let us review what we have seen concerning water and our relationship to it. It is a scientific fact that the human body is 75% water. So that this hydration that ***IT*** reshaped into makes skin elastic and a perfectly fitting cover that keeps us warm, prevents our tissues from drying out, and protects us against dirt and germs and the Sun's harmful rays. The skin's hydrated cells, along with fatty tissues, also cushion the body against injuries. Water is also present in our tears and our saliva, which act as antiseptics, to protect our eyes and mouth, which are very vulnerable to the environment. Many things affect our bodies all the time, such as friction, heat, and cold. Taking these daily assaults, we should be grateful that our bodies do more than just survive! So the skin ***IT*** reshaped into as ***ITS*** weight, which is so dependent upon water, is what allows us to exist and adapt to the world around us.

Now remember that ***IT*** made water as something that could absorb the most heat, more than any of the other substances ***IT*** reshaped into, so that we could exist as the human body that has a relatively constant body temperature of thirty-seven degrees centigrade (ninety-eight degrees Fahrenheit).

Now remember that it was ***IT*** that made our bodies' existence possible, and remember that ***IT*** accomplished this reshaping before ***IT*** could see through our eyes, and ***IT*** was ***IT*** that, knowing that temperatures would rise, placed another part of ***ITSELF*** as ***ITS*** weight in the brain area that could receive and understand the skin's signals detecting heat and respond through the part of the brain called the hypothalamus, which stimulates the sweat glands so that we can perspire. This way we are able to lower our body temperature using the water that ***IT*** reshaped into, to cool us down, because high temperatures will put us in a danger zone. We should be more than grateful that ***IT*** reshaped ***ITSELF*** into water, which absorbs the excess heat and evaporates it, thus cooling our body.

And when we encounter the opposite situation, ***IT*** also placed mechanisms that come into operation when our body temperature drops too much or too fast. Our skin helps raise the temperature again by contracting what ***IT*** exists as in the form of our skin's *erector pili* muscles that cause goose bumps. They help us stay active and warm.

I was thinking about ***IT*** as water and wondered if this may be the only planet that ***IT*** reshaped into with so much water, that we think of Earth as "the water planet" since water occupies seventy five percent of its surface! For some reason ***IT*** concentrated ***ITSELF*** as water in the Southern Hemisphere as salty liquid and in the North as clean, frozen water and as fresh running water in between as rivers and mineral springs.

Water is so important that ***IT*** also increases the amount of available water approximately two inches every two thousand years. ***IT*** sends this water from the upper atmosphere in the form of water balls, which may contain different kinds of life as ***ITSELF***. As far as we know, this planet is the only place where ***IT*** as water, exists naturally in 3 different states: liquid, solid, and gas.

Considering how ***IT,*** as water exists as a liquid, the next time you look at water as the oceans, lakes, rivers, snow, and even clouds, remember that the water you see on Earth in all three states is ***IT*** as it reshaped ***ITS*** weight into hydrogen and oxygen atoms that come together to form water.

Besides changing into liquid, solid, and gaseous states in the hydrologic cycle, water is also affected by other powerful planetary forces, that ***IT*** also exists as, such as solar heat, gravity, and the tides generated by the pull of the Moon and Sun, that ***IT*** also exists as. Remember that all of these forces exist only because ***IT*** reshaped into them.

Now imagine that you can observe the planet Earth from space: You would see it in continuous motion due to its rotation; as ***ITS*** weight and as ***ITS*** speed, which is what makes everything on Earth exist in a continuous motion. So we know that due to the Earth's rotation, ocean water moves in an Easterly direction. But don't forget that Easterly is just a name. In reality, ***IT*** has no East, West, North, or South, or up or down, or left or right.

Another important thing to notice is that as ***ITS*** weight, ***IT*** keeps ***ITS*** weight in different ways, as the elements that ***IT*** also reshaped into as chemical substances. So now, the next time you hear or read that our oceans contain a vast number of chemicals, remember that this warehouse of chemicals that came from run-offs into rivers, are totally ***IT*** as ***ITS*** weight or as the pure energy that exists as these chemicals.

The point is for you to remember that water, in its liquid form, is one of the ways ***IT*** used/uses ***ITS*** weight to reshape ***ITSELF*** as cells, so that ***IT*** could become tissues and then organs, so that ***IT*** can move, think, and reshape ***ITSELF*** further. I know this sounds strange, but remember that you are every atom that ***IT*** reshaped into and you will see this better if you see yourself as you really are, made of that pure energy, which is that stuff that we know cannot be created or destroyed.

Now let's look at ***IT*** as ***IT*** exists as a gas, which is where ***IT*** holds as ***ITS*** weight as atmospheric water. There are approximately 3,000 cubic miles of water as ***ITS*** weight, suspended over our heads!

Let me mention that in 2 of ***ITS*** ways of existing as water, that is, as a gas and as a liquid, we have given the title of the hydrologic cycle in which ***IT*** uses ***ITS*** weight as solar energy to evaporate ***ITSELF*** as ocean water back into the atmosphere, from where it then later returns ***ITSELF*** as rain.

IT then sends this rainwater that came from evaporating ***ITSELF*** as the water from the sea, back again as rain that will fall on land where it deposits nutrients like nitrogen, and the minute amount of salt that is needed, and then picks up the nutrients that exist on land, as ***ITSELF***, as minerals that are now running from rivers to the oceans, to feed ***ITSELF*** as the life that ***IT*** also exists as, as one of ***ITS*** possibilities of existing as life that needs these nutrients. In addition, approximately one sixth of the atmosphere's water penetrates the ground as rainwater and dew.

So that as ***ITS*** weight as rain, ***IT*** then saturates land as rain as humidity, where ***IT*** then takes this water and raises it as ***ITS*** weight, by using ***ITSELF*** as the solar heat that ***IT*** also exists as, through capillary action, that will bring ***ITS*** weight that exists as nutrients, that enter plants through their roots.

Now, since ***IT*** brings back this water through evaporation, using ***ITS*** weight as heat from the Sun, once this gaseous water vapor returns to the atmosphere, because it is lighter then water as weight, and in being lighter, some of the gases present in the atmosphere, such as nitrogen do not readily want to return to land again. What ***IT*** then does is that, as ***ITSELF*** as a lightning discharge which is nothing more than electricity, ***IT*** welds together the nitrogen, and other nutrients and oxygen molecules found in the atmosphere and sends them back with the rain, so as to supply plant life with fertilizers. We can all bear witness to this process, because we can see that when rain falls, plants grow faster because of the nitrogen, that was in gaseous form in the atmosphere, light in weight, was made to return to the soil again. We can even smell this nitrogen on our clean laundry when we set it to dry outdoors and rain falls on it, because as ***ITS*** weight as rainwater ***IT*** provides ***ITSELF*** as plant-life with nitrates that function as what we call fertilizers. Rain also contains sodium chloride (salt) which is present in sea water.

Now, ***IT*** also moves what ***IT*** exists as rainwater that contains organic components that entered the atmosphere by way of the water evaporating from the leaves and other surfaces of plants, in which ***IT*** also exists as the microflora and microfauna that exist as bacteria, spores, and plant seeds.

And since ninety percent of the Earth is made of five elements: oxygen, calcium, silicon, aluminum, and iron, ***IT*** uses these elements that ***IT*** exists as, to supply ***ITSELF*** as the plant life that exists on land and in the water, supplying ***ITSELF*** as the fruits and vegetables that ***IT*** exists as, that have the minerals, that are needed for what ***IT*** exists as the life forms that ***IT*** reshaped into on land and in the water, and ***IT*** also sends these nutrients back to ***ITSELF*** as the sea that ***IT*** exists as, so as to support ***ITS*** way of existing, as all the ways ***IT*** reshaped into as the many ways that ***IT*** as life exists as what we see as trillions of living things that exist on this planet, both on land, in the rivers and lakes, and under the sea.

Let me throw in a note here: When we say that water is becoming acidic, we should remember that rainwater has always been slightly acidic. We know that water is basically made of, 89% oxygen and 11% hydrogen by mass, in nature. But we rarely find pure water in this form.

I am sure you remember from science class in school that the water present on our planet today is the same water that existed at its birth. I believe it may be the same water, but the components of the water will not include the same contaminants. This is because of ***ITS*** reshaping. Changes will occur so that current contamination will not be the same contamination that exists in the future. In melted ice glaciers, we have found substances of our current water supply that were not there when that water had been frozen. There are core samples of glaciers that go back thousands and thousands of years that do not contain contaminants that exist today.

IT as a gas. IT gave ***ITSELF*** to us as our human essentials: such as oxygen and water as ***ITS*** weight, so let me start with water which plays such an important role in our lives. It is one of the three essentials required for our human existence. In order of importance, these essentials are first, oxygen, second, water, and third, fuel (food). Oxygen, our first necessity, is element #8; it has eight protons, and I will refer to oxygen as "number eight" to illustrate what happens when we breathe. We inhale number eight, along with other gases present in the atmosphere, but we exhale number six, carbon, which occurs in molecules containing one atom of carbon attached to two oxygen atoms: carbon dioxide. In order for the lungs to perform their function there is an exchange of gases. Hemoglobin is a key substance. When we inhale air, it captures oxygen molecules and takes them to the cells, and on it's way back to the lungs it carries carbon dioxide molecules to be released into the atmosphere when we exhale. The blood also removes carbon dioxide molecules from the cells. Element number six, carbon, which is a part of the carbon dioxide, is lighter than number eight and will then rise into the atmosphere where it will be absorbed by plants and trees and with the aid of the Sun.

There's an old saying about our second necessity: "water is the essence of life." I believe it should read "**water allows *ITSELF* to exist as life with mobility.**" Water consists of two hydrogen atoms and one atom of oxygen. Hydrogen atoms act as drivers. They attach themselves to oxygen, driving it to its destination. The hydrogen atoms that are a part of our red blood cells, for example, drive oxygen to our brains. This process could be called **C**onstant **O**xygen **D**emand **(C.O.D.).** A human body is 75% water, which as described above, is a combination of hydrogen and oxygen. A human life begins weighing anywhere from a few ounces a couple of months after conception to several pounds at birth as ***ITS*** weight. As a human life grows, we will continue to gain as ***ITS*** weight as water, along with other vitamins and minerals. Let us say that a human makes it to maturity and weighs a total of 100 pounds. This 100-pound human has 75 pounds of water. Since water is made of hydrogen and oxygen and each of these elements are gases, hydrogen, an explosive gas, and oxygen, a non-flammable gas, we could say that the 100-pound human is so-to-speak 75% gas, as ***ITS*** weight reshaped.

When this 100-pound human being dies, the body decays and only the 25% that comprises its skeleton remains; the gases return to the atmosphere and the vitamins and minerals leak into the soil. The skeleton consists of 18% carbon, 3% nitrogen, 3% calcium, and

1% phosphorous, all of which are elements made of atoms, which are 95% empty space. The skeleton is thus 95% empty space. From beginning to end we are 95% empty space.

The brain is also 85% gas and 95% empty space. The brain transmits information by converting energy into words that enable us to understand our thoughts. It is said that we only use an average of 10% of our brain capacity.

The human brain weighs approximately three pounds. In any case, these meager gaseous pounds believe themselves to be the almighty of everything that exists in the Universe. And this gas transfers thoughts into vibrations that convert into what we know as conversation.

The next time you look at a fellow human being or an animal, even if it's huge, such as an elephant, remember that you are actually looking at a combination of ***ITS*** weight as gases such as water vapor, and are 95% empty space as atoms. Here we see the 5% to 95% ratio ***IT*** uses to arrange the Universe: 5% matter and 95% empty space that exists as nothingness.

The next time you see or come in contact with water, remember that it was ***IT***, in ***ITS*** reshaping as ***ITS*** weight so as to become water, starting from the oceans, where ***IT*** then used ***ITS*** heat from the Sun to evaporate these ocean waters, that will take with them a minimal amount of salt, so as to drop this salt on where ***IT*** exists as land, so that this salt would then move into fruits and the flesh of animals that we eat, as the minimum amount of salt that our existence needs, for we do not really need the salt-shaker. We do not need to add salt to our daily diets, because the amount of salt that we need, as what ***IT*** placed in us as our requirement, ***IT*** supplies to us with this amount from regular water, and from the foods that we eat that already contain the amount that we need in order to exist with out having high blood pressure.

IT did all this so that we could get what is needed in order that we could have a body with the proper amount of salt, and so that we could exist as our water body, where ***IT*** placed ***ITSELF*** as our human brain, and as a heart that could pump the water as blood so as to deliver oxygen to every cell that we exist as, and as the oxygen that energizes our brains.

So I say thank you, to ***IT***, for having the know how to use ***ITS*** weight in reshaping to arrive at being a living cell, so that I could at least say thank you to ***IT***, for becoming a cell that exists in this place that ***IT*** reshaped into that we call Earth, (the water planet) and thank you for knowing just how much salt was needed in my body in order to enjoy your existence as life.

Observation . Now in my observations, I have noticed that as ***IT*** begins to use ***ITS*** weight as us and other living things, ***IT*** is adding more of ***ITS*** weight, as our organs, for as we start out as babies, our skin, heart, liver, lungs, and brain just keep gaining, as in adding more of ***ITS*** weight to these organs. At least that's the way it has been in my existence. No wonder I always have to be watching my weight.

⌘~~~~~~~~~~~~~~~~⌘⌘~~~~~~~~~~~~~~~~⌘

We can see ITS dual temperatures right here on this planet: The North and South Poles as ITS cold and the Equator as ITS heat

❁~~~~~~~~~~~~~~~~~~~~~~~~~~~~~~~~❁

Now let's see how *IT* exists as glaciers. We should remember that in spite of being a solid, ice is less dense than water in its liquid state; this is what enables ice to float. What gives us the illusion of density is that at four degrees below zero Centigrade (-4°C), glaciers are certainly rigid. Also, when water freezes, it expands as much as 9%, so that it occupies more space. This allows icebergs to float as they break off a glacier. This is an example that shows how ***IT*** uses dual temperatures: coldness to turn liquid water into ice and heat, to melt ice back into liquid water.

Scientists have learned that if these northern glaciers were to melt as a result of the heat that ***IT*** sends us as ***ITS*** weight from the Sun, the water levels of the Earth's oceans and seas would increase from 65 feet to 195 feet. We as humans see things like this as undesirable events, but one very important thing to always remember is that everything that exists is ***ITSELF***, as pure energy, or as religions would say, God created everything. It may be that ***IT*** needs to move ***ITS*** weight as glaciers into more available weight as water, which maybe needed by the ever growing human population, which needs 75% water to allow their bodies to exist, an as the amount of water that they will need to consume as drinking water, and as the water that plants and animals will need for themselves and also in order to supply our daily food intake.

So it maybe that ***IT*** is melting ***ITSELF*** as these glaciers in order to have more available, clean water for ***ITS*** reshaping into the above. One more thing that we should always remember is that ***IT*** will always be looking for all existing possibilities in which to reshape ***ITS*** weight.

⌘~~~~~~~~~~~~~~~~~~~~~~~~~~~~~~~~~~~~⌘

***** Water is what gives life mobility, but it is speed that directs it.*****

❁~~~~~~~~~~~~~~~~~~~~~~~~~~~~~~~~❁

My second book, Omnipresent, Volume II, is one of an even more astute level of understanding. As the reader, you will begin to gain a higher understanding of how ***IT*** began to reshape ***ITS*** heated-weight within ***ITS*** freezing-cold shell body that we call the empty Universe. As a reader, you will also begin to understand how time really exists as just a human invention, as a human convenience, and how ***IT*** exists as omnipresent, as in being in all places at the same moment, not subject to our mechanical time system. Here ***are some of the subjects that are covered in volume #2:***